Evolution – A dual Process

A new RNA
Elucidates Darwinian Evolution

By Andreas Polydorou BSc(Hons), PhD, FRSA

Dedicated to my wife Margarita whose continuous motivation has enabled me to proceed with the publication of this book after 30 years of deliberation.

© 2007 by Andreas Polydorou

Andreas@adm-computers.com

2nd Edition 2015
First Published 2007
ABC Publications London.

Review by S. Smith (Author, Scientific Reviewer)

….Scrumptious mind-fodder. Complex scientific theories are dished up in an exemplary easy-to-read manner, both in the language used and in the layout on the page. This is popular science at its best, his hypotheses argued with enthusiasm, his every supportive example being in itself highly informative and entertaining.

….Excellent. Reads like being given a series of talks by one's favourite professor. Not only does the author deliver up, in entertaining and diverting fashion, scientific history, he works into it the narrative of his own journey of discovery, confesses his trepidations when going against accepted opinion, thus involving us in his temerity and gets us willing him on. Where scientific terms are unavoidable he explains them succinctly in layman's terms. And, in acknowledging that much of life is yet a mystery, so open is he about his own doubts that the reader soon learns to trust his every assertion as a genuinely held belief and therefore worthy of consideration.

….this book could become as popular as Desmond Morris's 'Naked Ape'; or, given its prognostications, as influential as Gordon Rattray Taylor's 'Doomsday Book'.

Evolution – A dual Process

A new RNA Elucidates Darwinian Evolution

By Andreas Polydorou BSc(Hons), PhD, FRSA

Introduction		7
Chapter 1	Appearance of Life on Earth	19
Chapter 2	Unexplained observations concerning the Human Race	36
Chapter 3	A new Hypothesis	68
Chapter 4	Singularity Vs Duality	121
Chapter 5	Mammals	140
Chapter 6	Fishes and other sea creatures	164
Chapter 7	Reptiles and amphibians	185
Chapter 8	Birds	199
Chapter 9	Insects	214
Chapter 10	Plants and Flowers	230
Epilogue		247
Appendix 1		249
Appendix 2		258
Appendix 3		259

INTRODUCTION

In this book I present a Hypothesis that Evolution is a dual process: Ordinary Darwinian Evolution plus another process which forms part of my Hypothesis. This is based on a proposed new RNA.

This book considers various unexplained observed phenomena associated with Darwinian evolution and presents a new Hypothesis through which these problems are analysed and possible solutions provided. My Hypothesis does not reject Darwinian evolution. In fact I stress throughout that Darwinian evolution is necessary in the creation of new species and I explain in more detail as to how this is achieved. My Hypothesis simply complements Darwinian evolution in trying to resolve the difficulties that are encountered in the further development of a species towards perfection and enhancement of its chances for survival.

This is a very important issue because evolutionists have always assumed that there is only one single mechanism of evolution. They assume that the creation of a new species is effected by the same mechanism as the mechanism followed for the further development of the new species towards perfection.

I propose that the further development of the new species is based on an entirely different mechanism, which forms the basis of my book. This further development follows a definite course with a scientific explanation that is not based on randomness.

Furthermore reasons are given as to why it is scientifically impossible for randomness to provide this course of evolution. For example it contradicts the Second Law of Thermodynamics, which is one of the most fundamental Laws of Nature.

Darwin freely admitted that he could not provide an explanation to the fact that unlike all other land mammals, the human body is not covered by hair. Through my Hypothesis I try to give a reasonable scientific explanation as to how this characteristic has evolved. Similarly I try to give such explanations as to why the human penis has no bone in it, unlike all other mammals, and why humans normally have sex by facing each other, rather than through the posterior position adopted by virtually all other animals. Also why women are the only female creatures to experience orgasms?

The first four chapters of the book are dedicated to explaining my Hypothesis through established scientific principles involving a number of examples where Darwinian evolution simply fails to give an adequate explanation. The last six chapters deal with nearly two hundred instances where Darwinian evolution fails to give a satisfactory explanation. My Hypothesis can be used to explain all these examples. I should point out that the list of examples is nowhere near exhaustive. I am sure that many readers will come up with additional examples whilst reading through the pages.

It is imperative that I should draw attention to the fact that there is a big discrepancy between Darwin's original Theory and present day accepted Darwinian Theory. For example Darwin in the first Chapter of his first book "Origin of Species" says:

"Changed habits produce an inherited effect, and with animals the increased use or disuse of parts has a marked influence; thus in the domestic duck the bones of the wing weigh less and the bones of the leg weigh more, in proportion to the whole skeleton, than do the

same bones in the wild duck; and this may be safely attributed to the domestic duck flying less and walking more than its wild parents..... Not one of our domestic animals can be named which has not in some country drooping ears, and the view has been suggested that the drooping is due to the disuse of the muscles of the ear, from the animals being seldom much alarmed."**

Also he starts Chapter 5 with the following statement:

"I have hitherto sometimes spoken as if variations were due to chance. This, of course, is a wholly incorrect expression, but it serves to acknowledge plainly our ignorance of the cause of variation."

Unfortunately modern Darwinian Theory has rejected these premises as described by Darwin. And this rejection has been done without any scientific foundation – simply on inductive reasoning. What I am trying to do in this book is to provide scientific foundation to prove that there is some truth in Darwin's original assertion, as defined in the above paragraphs. I am not trying to say that everything that Darwin said was correct but that some of his ideas that have been rejected may have been rejected without full justification.

I suppose this makes me more Darwinian than the strict present day Darwinians. I regard myself as a revisionist of the Theory rather than an annihilator of it.

Obviously in my endeavour to provide a proof for these premises I have to use scientific principles unknown to Darwin. I have to refer to the structure and function of the cell especially the DNA and RNA.

Throughout my book when I refer to Darwinian Theory I refer to the modern adaptation and not to the original proposals of Darwin.

I have tried to ensure that all the examples I present in my book refer to characteristics that are genetically transferred from parent to offspring. This is sometimes not easy to define. There are many processes where it may appear that a characteristic is genetic but in reality this may not be so.

For example when we see Africans running in the jungle barefooted we get the impression that the soles of their feet have acquired a very tough skin which gives them similar protection to what other people get when wearing shoes. I have personal experience of this. When I was a kid I used to run around barefooted and my skin did indeed become so tough that even most thorns were broken under the pressure of my weight. If a city person walked barefooted in the fields that I was walking, the thorns would immediately go through the skin of their soles. But the main thing is that with toughened soles I, like many Africans, found it quite comfortable to walk about on any surface. Later on in my life when I started wearing shoes, the soles of my feet became soft just like any other city person. In other words this toughening of the soles is not a permanent characteristic. Therefore it is not genetic. The ability however for the soles to get toughened when required must be genetic because virtually all humans have this ability.

Let us look at another example which I consider to have genetic characteristics.

Most people in the world drink hot tea at least once a day. Humans are the only creatures that have the ability to withstand the temperature of tea, which can be as high as over 80 degrees Celsius. The mouth would normally be regarded as a sensitive area, much more sensitive than the external skin of man. Yet if water at that temperature were to be poured on someone's arms or legs, a terrible scalding would result in addition to the immediate and unbearable pain. But hot liquid intake by man cannot be that old, possibly three to four thousand years old. It is obvious that the ability of the mouth to withstand the high temperature has evolved not due to randomness but due to a self imposed pressure by humans to be able to enjoy a hot drink or a hot meal. If a person stopped drinking hot drinks for many years, that person could still enjoy a hot drink without any scalding to his mouth if he had a hot drink after many years. This is why this particular feature can be classified as genetic.

This can be extended to cover food or sweets which contain hot spices. Most people like them. Chimpanzees however detest the taste of menthol or rather the pain that it causes in their mouths. I personally gave such sweets to some chimpanzees whilst on a safari and my car was attacked soon after by them, damaging the wipers and the rubber seals of the front screen. They were boxing the front screen and the side windows so hard I thought they were going to break them down. The pain that the menthol caused to them must have been quite indicative of their reaction.

In fact humans have become accustomed to getting pleasure out of pain as clearly illustrated by the way that many enjoy some strong alcohol such as whisky and vodka. One could say that humans are the only creatures that display a masochistic attitude to life.

Let us look at another example.

The "blind spot" of the eye is due to the bunching of optical fibres at the back of the eye. One would have expected this bunching to be in the centre of the back of the eye on the line of symmetry from the iris. However if this was placed at this position then we would be unable to see at all. So the position of the bunching is off the line of symmetry. This required a very methodical design, one hardly achievable through randomness. I look at this example in further detail later.

This is precisely what my hypothesis tries to explain. I consider a number of examples that cannot be explained though Darwinian randomness. I also look into examples where Natural Selection could not possibly apply either because of time or numbers limitation or because of the survival rate being very high such as in the case of man. Without my hypothesis there is no explanation for these examples.

-.-.-.-.-.-.-.-.-.-

Darwin has said that even if one case is found that contradicts his theory of Natural Selection then the whole theory would have to be questioned and possibly collapse.

With my Hypothesis this is exactly the opposite. If only one of the examples I give here is proved to be correct then my Hypothesis stands.

It will be wrong to consider in isolation any one of the points that I mention in this book and decide that if you don't agree with my interpretation of one of the examples presented to reject the whole proposal made in this book. What I am proposing here is of great scientific significance. It could have far reaching consequences on our outlook on Biology in general and on Evolution in particular.

-.-.-.-.-.-.-.-.-.-

I would like to conclude the Introduction to my book with five important observations:

1. There are virtually no redundant organs or characteristics on any species alive today. Scientists spend a lot of time identifying the functions of newly observed organs of all creatures. They know that if an organ is there, it must have a function. Non-use of an organ due to changes in the environment etc. can lead to atrophy, but this is not very common.

2. Virtually all organs are perfectly designed on all species. There is hardly a single badly designed organ or characteristic on the two million species alive today. Such perfection is frightening for any design engineer. The design in some cases is so optimised that Engineers find it amazing. They feel that to reach such an optimised design will require the input of several thousand engineers and scientists working together through the world's most powerful computers. To mention just one example, the evaluation of stresses in a structure is normally done through Finite Element Analysis which requires enormous computing power – such as that offered usually by the Gray main frame computers. Even that can sometimes take many hours or even weeks of computations. Such technology was used and is being used in the design of the take-off platform of space rockets and the space shuttle. The web-shaped design of the outer sections of these platforms is amazingly similar to that adopted by some trees such as the Tetrameles nudiflora in the Vietnamese jungle. Nature provided this amazing solution without the use of computers or Scientists of cumulative experience equivalent to millions of years. These massive trees can exceed 50 metres in height and their web based structure close to the ground keeps them robust and sturdy.

3. There has been very restricted evolution for millions of years. Only a minute proportion of species have changed during the last few million years. Most of them have not evolved in any way for many tens or hundreds of millions of years – jelly fish (600 million), Platypus (300 million), dragon fly (300 million), horseshoe crab (400 million), army ants

(100 million) etc. Those species that have changed have done so under environmental pressure and in many cases present good examples of rapid evolution. In other words special characteristics have evolved in regions or under conditions where they were desperately required as illustrated by the following:

a. Competition for survival:
    1. Giraffe – the classical example first discussed by Lamarck
    2. Cat and mouse situation such as the case of the bat and the moth described later on in this book
b. Movement from sea to land – and vice versa
c. Movement from land to air – insects, birds, bats
d. Movement of species to very cold regions such as the Arctic and Antarctic. The Arctic bear, Eskimo and ice fish are such examples.
e. Movement to very hot regions such as to some deserts. The Camel and the Cyclorana frog of the Australian desert are typical examples.
f. Movement to very low sea depths where no light exists – several species have evolved with unique characteristics. Over 90% of these creatures have evolved their own light source using bio-luminescence.
g. Movement to high altitudes with little Oxygen – Langur monkeys, Sherpa
h. Isolation – Galapagos Islands, marine iguana
i. Intuition for survival - army ants, bolas spider
j. Man – self imposed pressure by the brain. This is the best example where Natural selection does not apply as there is normally a 97% chance of survival to adulthood.

4. There is strong evidence for rapid evolution especially in the case of man. The cases of the evolution of the special characteristics of the Eskimo and the Sherpa mentioned above as well as the evolution of the Mongolian eye are very good examples of rapid evolution concerning humans. Other good examples are the domestication of the Abyssinian cat and the evolution of the dingo from domestic dogs or cross breeding with wolves. I should also cite my own work on ants which showed that domestic ants have acquired the ability to climb up a drinking glass whilst ants in the jungle are unable to achieve that. All these are examined within this book later on in greater detail.

5. Nature has come up with virtually identical solutions for entirely different species:

    a. Vampire bats and leeches live off the blood of their victims. They both have razor sharp teeth to pierce the skin of their victims. They both developed an anaesthetic that kills the pain as they suck the blood. They also both developed an anticoagulant agent in their saliva that ensures that the blood continues flowing whilst they are sucking without clotting.

    b. Ant eaters around the world have developed a very long snout and a long sticky tongue with which they pick up the ants and termites from their nests. Such ant eaters include the giant ant eater and the armadillo of South America, the pangolin in Africa and Asia, some marsupials in Australia, etc. Even birds that have taken to ant eating such as the woodpecker and the wryneck have also developed long beaks with very long sticky tongues. One other common feature of these creatures is that they have all lost their teeth due to atrophy.

c. Negroes in Africa and the Aborigines in Australia have evolved virtually identical skin colour. This is due to the melanin created as a result of the sun's radiation. Melanin increases the rate of absorption of the useful part of the sun's radiation, which in turn provides additional protection to the body. Equally important is the fact that the tanned body reflects the harmful Ultra Violet radiation.

-.-.-.-.-.-.-.-.-.-

Please note that this is not a textbook. It is aimed at mass readership, even though it can also be of help to students of science.

I have been planning this book for over thirty years. I started writing it on different occasions in different forms. I could not make up my mind as to which was the best way to present it. The first form followed one where it started with a detailed study of the cell and the DNA structure with numerous diagrams, etc. It then followed a detailed analysis of Lamarckism, Darwinism and Mendelian Genetics. It was too heavy for most people. I then started changing it, making it simpler and simpler until it reached the present form.

There have been several scientists who tried to bring the subject of the revision of Darwinian Theory for discussion, but unfortunately the strict Darwinians have always treated them with contempt. It took a lot of courage to even start to write this book because I knew how it was going to be treated by most Darwinians.

Because of this, I have to reiterate more than once that **my hypothesis, does not contradict Darwinian Theory, it simply complements and augments it.**

Einstein said that a good theory has to be elegant. Darwinian Theory on its own lags elegance. It is ugly; it is inept; it is clumsy; it is disjointed; it is incomplete. Many free-thinking Scientists reject it. It contradicts basic scientific principles such as the Second Law of Thermodynamics. My hypothesis offers elegance, style, finesse, unity and completeness. My hypothesis is simply beautiful.

CHAPTER 1

APPEARANCE OF LIFE ON EARTH

Life appeared on Earth some 3.3 billion years ago in the form of micro-organisms. The means by which life appeared on Earth is a matter of conjecture, but Scientists have come up with fairly interesting and sometimes fairly convincing evidence as to its origins. It is worth considering these origins before we proceed to the discussion of how life has evolved through various stages to reach the present state of millions of comparatively advanced animals and plants.

I will start by presenting a brief introduction on scientific views relating to the position of the Earth compared to the rest of the Universe and how life may have first appeared on Earth.

There are 9 planets orbiting around the sun, which together with their own satellites, form the solar system. Earth is the third planet from the sun and is the only planet with life on it. The two planets closest to Earth are Venus and Mars. Venus is closer to the sun than Earth and is far too hot for life to appear on it, whilst Mars is further out and far too cold for life to have developed there.

Earth at 150 million kilometres from the sun seems to be at the right distance and with a radius of 6,370 km seems to be the right size to have the correct temperature as well as the correct atmosphere to allow life to exist on it.

Earth is estimated to be about 4.5 billion years old, whilst the sun is estimated to be about 4.7 billion years old. The Universe is assumed to be about 13.8 billion years old.

The latest theories assume that the planets around the sun were formed due to collapsing gas clouds, which existed around the sun after its formation.

The formation of planets around the sun through collapsing clouds of gas enhances the view that billions of other stars in our Galaxy may also have planets around them. In 1995 observations confirmed the existence of such planets around some of the stars. In 1999 multi planet systems were also observed. Most scientists now take it for granted that there must be billions of planets in our own Galaxy alone.

One could arrive at the conclusion that amongst these planetary systems, there must be a high proportion, where at least one of the planets around a star must be at the right distance and of the right size to be able to sustain life on it. Two of the extra-solar planets observed recently, were found to be of about the same size as Saturn, only about 3 times bigger than the Earth's radius. Considering the fact that there are about 100 billion stars in our Galaxy, one would realise that the possibility of life in probably millions or even billions of planets, is reasonably high.

It is estimated that there are about 500 billion Galaxies in the Universe, some of them much bigger than our own Galaxy. When one looks at these enormous numbers one stops thinking in terms of whether there is life somewhere on other planets, in other planetary systems or in other galaxies, and starts wondering more as to what form such life might have taken.

Using a formula developed by Drake, some astronomers suggest that the number of civilizations in the Earth's galaxy alone may range from a thousand to a million. Notice that here the astronomers talk about civilizations on other planets, not just planets that may have some sort of life on them.

According to the big bang theory, the Universe originated between 13 to 15 billion years ago. There is evidence that all galaxies are moving away from one another and leads to the conclusion that the universe is expanding. Most of the planets we mentioned above would therefore be further out from the centre of the Universe, as defined by the big bang theory, and would therefore be much older than our solar system.

These planets that are billions of years older than Earth could therefore have civilizations that are infinitely more advanced than our own civilization. If you look at the technical and scientific discoveries that have taken place on Earth during the last few thousand years, you could imagine what it could be like in a million years from now on Earth. As these civilizations may be billions of years older than ours, they may make our technological achievements look pretty prehistoric compared to theirs.

-.-.-.-.-.-.-.-.-.-.-

Having looked at the position of the Earth relative to the rest of the Universe, we would now continue with examining the conditions that prevailed on Earth soon after it was formed some 4.5 billion years ago.

The first traces of life on Earth are of micro-organism fossils dated some 3.3 billion years ago. Thus it took about 700 million years for life to appear on Earth after its formation. But how did the first traces of life appear on Earth?

The Swedish scientist Arrhenius proposed at the beginning of the twentieth century that life on Earth originated from living spores driven through space by the pressure of light. He named this idea the "panspermia" theory. Some meteorites that collided with Earth have been found to contain traces of molecules, such as amino acids that are the building blocks of protein. Scientists have also observed in outer space the presence of such molecules as water, ammonia, methyl alcohol, formaldehyde, etc. which are also examples of building blocks of protein. So it is quite possible that the first traces of life on Earth came from outer space. After all, as we saw above, the chances of life on other planets are extremely high.

As the complex organic compounds in interstellar space detected by radio astronomers include some molecules that may be considered the precursors of life processes, scientists have argued that there is a high possibility that the chemical reactions leading to life are widespread in the universe.

The prevailing present-day thought however, is that life formed on Earth in our own atmosphere through the action of Ultra Violet radiation or electrical arcing on the molecules that existed on the Earth's atmosphere at that time. The atmosphere at that time contained simple molecules such as hydrogen, carbon dioxide, methane, ammonia, etc. It has been suggested that the action of Ultra Violet radiation and lightning flashes could cause these simple molecules to form into larger molecules.

There was no oxygen in the atmosphere; hence there was no ozone. Therefore Ultra Violet radiation, which is mainly absorbed by ozone, could reach these molecules without previously losing any of its energy.

This was actually tried in a typical laboratory simulation by scientists, who managed to produce Hydrogen cyanide and aldehydes. Further high voltage sparking produced more complex molecules including four amino acids – glycine, alanine, aspartic acid and glutamic acid. All these four amino acids are frequently found in many proteins and in nucleic acids, which are the two essential ingredients found in every living organism. Researchers have also managed to formulate large molecules through polymerization of the single components.

In 1999 American researchers carried out tests on aromatic hydrocarbons, the most abundant organic molecules in the universe, exposing them to the conditions they would have encountered on the early Earth some 3.3 billion years ago, if they had been brought here by a meteorite impact. Several organic chemicals were created by the process—including quinones, aromatic ketones, alcohols, and ethers—that would have been essential building blocks for amino acids, from which life on Earth evolved.

Scientists assume that after these substances were formed in the atmosphere, they dissolved in the seas, thus forming a thin soup, which in time became more and more concentrated. It is thought that after the formation of this primeval soup, proteins and nucleic acids started to develop.

When you get a solution of such substances in water, droplets will form through surface tension, which will contain these substances at a fairly high concentration, and an enveloping surface membrane will be created round them. These droplets tend to increase in size by continually absorbing more of the dissolved substances in the water, until they reach a certain critical size, at which point they start to break up into smaller droplets. The critical size at which the droplets start breaking up depends on the constituents of the droplets, the temperature and the surface tension.

This mechanism described here is similar to the division of a cell and the production of two cells from one. It is thought that this process of building up droplets with high concentration of the correct substances, after hundreds of millions of years, led to the formation of the first cells.

It is essential to emphasise that there is no unanimity of opinion concerning the actual process by which the first cells appeared on Earth, and from which the first micro-organisms developed. The fact that there is no unanimity of opinion is not relevant as far as we are concerned. Whether the origin of life was terrestrial or extra-terrestrial is immaterial for the present discussion. This is not the main topic concerned within this book.

This is however of great importance from another point of view. If the origin of life was extra-terrestrial and it came to Earth via life bearing comets or meteorites, or they arrived on Earth directly through the high concentration of organic molecules that have been detected in space, that means that these processes could also have "seeded" other planets in other planetary systems. This would make the basis of the structure of all the cells universally identical. We should not forget the fact that, the DNA structure of all cells from the smallest micro-organism to humans is identical in every respect.

If this is the case viz. the DNA of all the cells is universally identical, then the life on other planets could be very similar to life on Earth. The popular image of strange looking creatures from outer space may be further from the truth than we would like to believe.

-.-.-.-.-.-.-.-.-.-.-

As we mentioned above, the first life that appeared on Earth some 3.3 billion years ago was in the form of micro-organisms. It took another 1.6 billion years for the next form of life to evolve. This evolved life was in the form of algae, which can be classified as the most primitive floral life, which can be dated at around 1.7 billion years ago.

There is then a big gap of any evidence of further life on Earth for a period of over one billion years. The next evidence of further life on Earth is of relatively highly organised forms of life such as sponges, shrimps, clams, starfish, worms, etc. This evidence suddenly becomes apparent in fossils dated some 600 million years ago.

The millions or billions of generations required to evolve from the single cell algae and micro-organisms to the highly structured forms of one billion years later seem to be missing. There is simply no evidence of their existence.

Darwin admitted: "To the question why we do not find rich fossilliferous deposits belonging to these assumed earlier periods prior to the Cambrian period, I can give no satisfactory answer."

This is one of the basic missing links in the history of evolution. This is one of the issues that my hypothesis will try to resolve, if indeed there was any life during this period apart from algae.

The first bacteria started producing oxygen as a waste product. This was the basis of the formation of the atmosphere, as we know it today. The biggest source of oxygen however came as the by-product of photosynthesis. The oxygen content of the atmosphere continued to rise. This started the formation of the ozone layer in the upper layers of the atmosphere, which started absorbing the Ultra Violet radiation from the sun.

It is possible that life mushroomed at around 600 million years ago because the level of oxygen reached such levels that it could benefit life formation. The creation of the ozone layer which reduced the level of Ultra Violet radiation reaching the surface of the earth helped enormously in the sustaining of life forms. So returning to the question raised by Darwin above it is quite possible that the reason that there is no evidence of any other life form prior to the Cambrian period is because there wasn't any life form apart from algae.

After the single cell organisms, multi cell organisms followed. One of the first multi cellular organisms to appear was the jellyfish, which still exists today in its original form.

It is nearly completely transparent because 99% of it is composed of water. Jellyfish has a single polyp and can swim gently through the sea by using its tentacles, which are also used to drag food into its digestive system. They do not possess any sense organs such as eyes or ears. They also do not possess a brain. They simply have a number of nerves that control its movement. The jellyfish can also sting its enemies through its tentacles.

The next stage of development of the jellyfish was possibly the Portuguese-man-of-war, which exists today in its original form 600 million years later. This organism, unlike the jellyfish, which possesses only one polyp, is made up of a few hundred polyps. The Portuguese-man-of-war is a much more advanced organism and it has developed a gas filled float that floats above the water surface and acts as a sail. Below the surface, hung the polyps. Some of the polyps are used for catching food, others for reproduction, etc. The polyps whose function is to catch food can have very long tentacles many metres in length and they possess numerous poisonous threads.

-.-.-.-.-.-.-.-.-.-

Scientists use fossils to gather information about the past. Sometimes one has to admire the conclusions reached through very limited data. It is one of those areas of Science that one has to use a lot of intuition and personal judgement to arrive at a conclusion.

Most evolutionists present evidence based on a few bones and parts of skulls dug out from the past. They seem to reach amazing conclusions from scant evidence. Most people do not realise how little information these people have in their hands. When they make a find of part of a skull or a bone from a body, they try to piece together all sorts of judgemental conclusions to create a theory concerning this find. Occasionally a find may consist of only a tooth or a small part of a bone. Frequently there is no unanimity of opinion concerning finds of these skulls or bones and the arguments and counter arguments can carry on for years.

What is more important is that frequently they can be wrong.

The most obvious example where many archaeologists were fooled is the famous case of the Piltdown man. In 1911 a discovery was made in Piltdown (Sussex, England) of part of the left half of a human skull, together with part of the jaw and two molar teeth. One of the most famous anthropologists of his time Sir Arthur Smith Woodward used the fossils to reconstruct an extinct genus of mankind "Eoananthropus", the dawn man. It was heralded as one of the most important discoveries of all time concerning the origin of man. It was thought that it was as important as the discovery of Pithecanthropus of about one million years old.

It was not until 1953 that it was demonstrated that this skull was deliberately implanted there to discredit some of the scientists involved in the case.

It was proved that the skull was relatively recent and the jaw actually belonged to a chimpanzee. This jaw was artificially coloured to look similar to the colour of the other bones of the find. It was one of the biggest scientific hoaxes of all time.

This example is not quoted here to discredit scientists of any particular field but to show that anybody can be wrong including some famous scientists, especially when faced with scant and incomplete evidence.

Let us look at another example where scientists once again arrived at the wrong conclusions. The coelacanth is a fish up to 150 cm in length that lived in lakes, swamps, seas and oceans. Many fossils were found around the world which dated from about 350 million years ago to about 70 million years ago. All the authorities agreed that this fish disappeared from earth 70 million years ago.

In December 1938 however a trawler in South Africa caught a coelacanth in its nets. This created a lot of excitement amongst many scientists, and they thought that this was a unique capture.

However, in 1997 a marine biologist whilst holidaying on the Indonesian island of Celebes, he spotted coelacanths in the local fish market. It was later found that coelacanths have been used by the inhabitants of the Comoros Islands as part of their normal diet, for many years.

This example clearly illustrates how scientists can be wrong because of lack of evidence.

-.-.-.-.-.-.-.-.-.-

Having looked at the various accepted possibilities as to how life appeared on Earth, some recent finds could possibly throw doubt on all these thoughts and hypotheses. All the above hypotheses are based on the assumption that the sun is the only source of energy responsible for the creation and continued existence of life on Earth. This is all based on the fact that the sun is responsible for supplying all the energy required by plants through photosynthesis. Animals directly or indirectly depend for their survival upon plants.

In 1977 however some findings threw doubts on this assertion. Geologists during that year discovered a hydrothermal vent in the Pacific Ocean 2,000 metres below the surface of the sea. What was astonishing about this thermal vent was the fact that there was an abundance of life around it, which nobody expected to exist. This community of life appeared to be independent of other life in the ocean. The conditions around this vent were such that most scientists assumed that life could not possibly exist under such conditions. They soon however discovered that this life was absolutely dependant on the hydrothermal vent. The creatures they found were there simply because the vent was there.

The hydrothermal vent was produced from molten rocks coming out from deep within the Earth, like a live volcano. The hot rocks superheat the water causing it to rise upward towards the surface at high speed. This is the hydrothermal vent. The temperature of this water can be up to 400 degrees Celsius due to the high pressure at these depths. It is also loaded with many minerals some of which create highly toxic gases. This is a very inhospitable environment, yet it abounds with life.

One would not expect to find any life at all at these great depths; yet around the hydrothermal vents, there is a greater variety of living creatures than one finds in the coral reefs or in a tropical forest. The numbers of these creatures are also staggering. The vast majority of them had not been seen before; they were simply different species of animals. There were tubeworms, clams, mussels, shrimps, sea anemones, crabs, fish, and others.

The main question that everybody was asking was as to how these creatures got the energy to survive in this environment where there was no sun light at all. There was no way that they could survive on the small quantities of food that came down from the upper levels. The problem became more complicated as they discovered that the first specimens of tubeworm that were brought up to the surface had no mouth, gut or anus.

Further investigations showed that over half the weight of this tubeworm was made up of chemosynthetic bacteria. As they found out, these bacteria are using the energy in hydrogen sulphide molecules that are highly abundant in the vent environment. The energy provided by the hydrogen sulphide helped the tubeworm to create organic matter in the same way that plants use the sun's energy.

For most animals, hydrogen sulphide is a toxic molecule. The hydrothermal vent species however found a way to overcome the toxic effects of hydrogen sulphide and just utilise its energy. Fish has been found recently that survive entirely on hot Sulphur.

A new similar hydrothermal vent was later discovered in the Atlantic Ocean. The amazing thing about this second find was that even though all life was of the same nature viz. dependent on hydrogen sulphide and chemo synthetic bacteria, the Atlantic vent was mainly inhabited by millions of shrimps.

-.-.-.-.-.-.-.-.-.-

In 1984 it was discovered, that similar colonies to those of the hydrothermal vents existed in the Gulf of Mexico. However there are no hydrothermal vents in the Gulf of Mexico. Also the species they found were different to those of the hydrothermal vents.

After a lot of research it was discovered that these animals were also dependent on chemosynthetic bacteria. This time however the molecules used were not of hydrogen sulphide but of methane which was bubbling up from the bottom of the sea.

These new finds have been called the cold vents as opposed to the hydrothermal vents.

-.-.-.-.-.-.-.-.-.-

These findings as far as I am concerned offer new possibilities in terms of hypotheses concerning the origin of life on Earth. It is true that scientists involved in the evaluation of these new findings of both the cold vents and the hydrothermal vents, are treating the new species found as having evolved from the species in the higher levels of the ocean.

However the question I am posing is "Could it be the other way round?"

No proof, no evidence; just a possibility!

-.-.-.-.-.-.-.-.-.-

At this stage I would like to pose two further important questions:

1. Is it possible that such life as found in the cold vents of the Gulf of Mexico, exists in other parts of the world where there are sources of methane such as for example in the Middle East?
2. Is it possible that other types of life might exist in other parts of the world, which are dependent on other molecules apart from hydrogen sulphide and methane? Simply molecules that can provide energy for life forms to exist.

-.-.-.-.-.-.-.-.-.-

Unfortunately I have to finish this introduction with a pessimistic outlook concerning the future of life on Earth. Due to the fact that some countries have not agreed to adopt practices to ensure that no further damage is caused to the ozone layer, many people are convinced that the ozone layer will soon be reduced to such levels that they have started predicting the catastrophe that will follow this regrettable situation. It is well known that in some countries like Australia, which is closer to the Southern Polar region, the effects of sunbathing and the possibility of cancer of the skin are much more serious than in any other country in the Northern Hemisphere. Scientists are talking about changes in climatic conditions, an increase in the average temperature on Earth, the melting of the polar ice, the increase of the ocean levels, the covering of large expanses of land by the sea, and so on.

I would like to add another source of concern which I have not seen being quoted anywhere else.

Due to the fact that the ozone levels will be decreased, the amount of Ultra Violet radiation reaching the upper strata of the atmosphere as well as the surface of the Earth will be enormously increased. This increase in Ultra Violet radiation will consequently increase the chances of interactions between various molecules as happened 3.3 billion years ago. With one big difference! Today we have infinitely more molecules in the atmosphere with which such interactions can take place. The results can be completely unpredictable. New molecules can be created with high toxicity or even worse, unknown viruses may be created. Because the number of molecules present in the atmosphere is infinitely greater, and because the variety of molecules is much higher, the chances of these catastrophic events taking place are appreciably enhanced. In fact these chances can be considered as realistic. Toxic molecules or unknown viruses could start appearing at any time.

During the 80's Scientists were baffled when they discovered AIDS. They could not understand how the HIV virus responsible for AIDS appeared suddenly. Many came up with various proposals as to its origin but even today there is no universal agreement as to its origin. One theory that seems to have acquired wide acceptance is that the virus was the outcome of sexual activity between man and some of the apes.

But if that is the case there is one question that has obviously not been addressed correctly. One would assume that if such a practice does indeed take place between man and the apes today, it couldn't be something that started recently. One has to assume that the sexual appetite of man in Africa has not suddenly increased to such levels that he had to turn to the apes for sexual satisfaction. Surely one has to assume that such practices go back a long way into History. But if that is the case, why didn't the HIV virus appear before?

Is it possible (and please note that I am only posing a question rather than a dogmatic statement) that the HIV virus is the outcome of new viruses created due to the gaps created in the Ozone layer as discussed above?

At the turn of the new millennium we have had the emergence of SARS, the killer disease that has the symptoms of the ordinary influenza. Nobody has as yet given a satisfactory explanation as to its origin. Again, is it possible that its origin lies in the gap of the Ozone layer in the atmosphere?

At the moment the thinning of the ozone layer has mainly taken place over the polar region where the atmosphere is relatively clean. So the consequences have not been that shattering. But as the thinning of the ozone layer progresses to other parts of the world where the abundance of industrial waste molecules is much higher and also the size of the molecules is much bigger, the chances of interaction might increase by a phenomenal amount.

Scientists have recently been baffled by the Asian haze, a thick brown cloud of pollution covering an area of about 10 million square kilometres. It rises up to three kilometres into the Earth's atmosphere. This was formed over the Indian Ocean, and it is thought that it is contributing to acid rain. It is also thought that it is reducing the amount of sunlight reaching the seawater. The affected area includes the Arabian Sea, most of the northern Indian Ocean, and part of the Bay of Bengal. This haze is a mixture of pollutants consisting of tiny particles of soot, chemicals and carbon monoxide from vehicle and industrial emissions and could cause absolute devastation on the region's climate and marine life.

The reason that scientists are baffled is because they simply did not expect this haze to be at the position that it is and also they did not expect it to be of such magnitude. It is thought that this cloud is the result of industrial and other emissions of countries such as China, India and others in South East Asia.

If this haze is out of the expected area by several thousand kilometres, probably due to various weather conditions, what could stop it travelling even further South towards the South Polar Region? If this were to happen, the interaction of Ultra Violet radiation with this high concentration of industrial molecules could be absolutely unpredictable and possibly devastating.

-.-.-.-.-.-.-.-.-.-

There is however one other threat. This is the possibility of development of new and unpredictable DNA structures. Let us face it, if 3.3 billion years ago the high concentration of Ultra Violet radiation managed to create all life that we have here on Earth today, what guarantees can we have, that new high levels of Ultra Violet radiation will not produce a different type of life form, that could even be the "anti life" of present life on Earth?

One could compare this with the collisions of matter and antimatter in space with the complete annihilation of both matter and antimatter and the emission of enormous quantities of energy. One could cite the famous meteorite that fell in Siberia on the 30th of June 1908. It caused a lot of destruction but apparently left no trace of itself, leading to speculation that it was a meteorite composed not of matter but of antimatter.

In his introduction to the 1979 edition of Darwin's Origin of Species, the world famous authority Richard E Leakey writes, "As oxygen built up in the atmosphere a layer of ozone developed above the oxygen, and this now screens out most of the ultra violet radiation from the sun. For these reasons, among others, a living organism could not arise *de novo* on earth today".

When Richard E Leakey wrote these words in 1979 there was no threat to the ozone layer. So he wisely said that new life forms could not be created. But he said that the main reason stopping new life forms being created is the ozone layer. So indirectly he agrees that in the absence of the ozone layer, the creation of new life forms is something that presents a possibility.

CHAPTER 2

UNEXPLAINED OBSERVATIONS CONCERNING THE HUMAN RACE

Before I start with outlining weaknesses and unexplained observations, I have to reiterate that I firmly believe that Darwinian Theory adequately defines most of the observed phenomena in Evolution. I have to make this statement many times because in the past, revisionists of the Theory have been cast as anti-Darwinians, something that I am not.

I simply agree with Darwin who said that his Theory does not provide adequate explanation especially in the detail of little used parts or characteristics which are not absolutely necessary for survival – simply make life more tolerable.

In this Chapter I will try to outline a few examples, some of them concerning the Human race, where Darwinian evolution does not provide sufficient or satisfactory interpretation. I will not try to present my thoughts as to how these difficulties can be resolved until the next Chapter.

-.-.-.-.-.-.-.-.-.-

The first person who tried to lay the scientific foundations of evolution was the French scientist Jean Baptiste Lamarck at the beginning of the 19th Century.

Taking the giraffe as an example, Lamarck suggested that the evolution of the long neck of the giraffe was the outcome of continuous exertion by its predecessors to reach the leaves in the higher branches of trees. Each generation succeeded in producing a slight increase in the length of the neck of the giraffe. After many generations the giraffe eventually acquired its very long neck.

Thus Lamarck advocated the view that the environment influenced organ development. He furthermore said that the development of new organs was based on the degree of requirement of these organs by the animal. He also suggested that change in body structure was based on use and disuse of parts. Finally he proclaimed that any acquired characteristics during the lifetime of a creature are inherited by its offspring.

Darwin, a few decades later reached the conclusion that Lamarck's explanation was wrong. He thought that animals and plants acquired new characteristics through random factors whose origin he was unable to describe, as at his time there was not enough knowledge concerning the elaborate details of the cell structure. Furthermore he suggested that the number of offspring of plants or animals brought into being is far greater than those that survive to sexual maturity. There is therefore always a battle for survival. Darwin argued that it was the fittest that survived. The others perished before reaching sexual maturity. Finally he concluded that the acquired characteristics that benefited animals and plants in their struggle for survival were inherited by the next generation. This he called the law of Natural Selection. Later this phenomenon was also referred to as Survival of the fittest.

Just before Darwin presented his theory at the Linnean Society he received a letter from Alfred Wallace who lived in the Far East. Darwin was amazed to see in this letter that Wallace had independently discovered Natural

Selection. They agreed that the paper should be presented jointly and it was agreed that it was their joint discovery. As Darwin lived in England he was much more involved in various lectures and publications and slowly the name of the theory evolved to "Darwinian theory".

There is however one important aspect of the joint theory. Wallace did not accept that their theory applied to the evolution of man. His argument was that there was no way that randomness and Natural Selection could possibly create such an advanced brain whose abilities were obviously much higher than man required for his survival. He wrote about it extensively expressing his view much to the dislike of Darwin who preferred to keep quiet about the particular issue of the human brain, as it appeared that he had a few reservations himself.

The main points to bear in mind concerning Darwin's theory of evolution are:

1. New characteristics are acquired through randomness
2. The number of offspring brought into being, is far greater than the number that survive to sexual maturity.
3. Therefore only those that survive can pass on acquired characteristics to their offspring.

Even if we forget for a moment the aspect of randomness, it is obvious that Natural Selection cannot apply in the case of man as the percentage of offspring that reach sexual maturity is so high that it does not allow for Natural Selection. This was precisely Wallace's argument with which Darwin agreed but he initially preferred to keep quiet about. Later however Darwin mentioned various issues that puzzled him concerning his own Theory. He was too honest a man to keep quiet about such issues. It is said that even on his death bed he was saying repeatedly that the blacksmith's son must have acquired his big muscles from his father.

Darwin considered the giraffe to illustrate the difference between his Theory and that of Lamarck. Darwin argued that due to randomness, some giraffes were born with longer necks than others. Those with the longer necks had a higher chance of survival because they could reach the leaves on the higher branches of the trees. In a period when there were not enough leaves on the trees due to a water drought or other reasons, the giraffes with the shorter necks could not reach the leaves on the higher branches and thus perished.

Darwin's Theory was soon accepted and replaced Lamarck's Theory. Since that time, there have been many exponents of Darwin's Theory and very few people tried to argue against it, as the evidence in support for it was so overwhelming.

In fact on a few occasions that people tried to express an opinion on some observations that could not be explained through Darwin's theory, their comments have been pushed aside as heresy. Darwin's theory is now as well established in scientific circles as Pythagoras' Theorem or Archimedes' Principle.

-.-.-.-.-.-.-.-.-.-

August Weismann towards the end of the 19th Century was the scientist that threw the final stone into Lamarckism to destroy it once and for all. Weismann presented a brilliant case but it was not based on any experimental work. The few experiments that he carried out himself, such as the cutting the tails of mice can go down as candidates for the most unscientific in History. I find it incredible that he carried out these experiments. I am sure that most people know that if a person loses a limb and then becomes the parent of a child then the child would have his or her limbs intact. In fact Darwin stressed this in his writings. But this is Science I am afraid!!! It can at times be as bad as or even worse than inconsiderate politics or ruthless marketing.

-.-.-.-.-.-.-.-.-.-

Frequently in this book I stress the fact that many aspects of what I am saying today was actually said by Darwin 150 years ago.

So what is new about my Hypothesis?

The new element that I am introducing is duality, viz. that there are two processes involved in evolution. Darwin believed that the creation of a new species and the subsequent evolution of that species towards further improvement were based on one and only one process. I am saying that these two are completely different processes. The mutations that create a new species are random. The mutations that improve the species are not random. This is my Hypothesis.

This duality is the basic reason why there has been so much disagreement amongst so many scientists – as they were trying to find or prove the wrong process. They were never looking for two separate processes.

I will try and present further on in this book a possible way that such a system works based on DNA-RNA interaction but even if that is eventually proved to be without foundation, my basic Hypothesis would still be valid. We need to establish the principle through which such a dual process evolution is based.

-.-.-.-.-.-.-.-.-.-

In this book I will try to present some examples where Darwin's Theory cannot by itself provide a satisfactory explanation. It has taken me a lot of courage, and more than three decades of deliberation to reach this stage of publishing my thoughts, because I know that some people that will not bother to read beyond this paragraph will immediately brand me as a heretic.

I should emphasise once again, that I am not dismissing Darwin's Theory. In fact I believe that in the majority of cases it provides a perfect explanation of the observations, concerning the origin of species – which incidentally was the title of Darwin's first book.

What I am proposing in this book is that in addition to Darwin's Theory of evolution due to randomness, there are other factors that can influence evolution, especially in the case where Darwin's Theory cannot explain how such changes can take place in a very small interval of time.

I will like to present a typical example where Darwin's Theory for natural selection provides an obvious explanation. Male deer during the mating season fight for the right to breed with a female deer. If one male deer was born without antlers, it could not possibly win the fight when faced with a male deer with antlers. As it cannot win the fight it stands no chance of mating with a female. Therefore the trait of no antlers cannot be inherited. This is a typical example where Natural selection is obvious.

Let us consider another example where Darwin's Natural selection provides adequate explanation. Soon after the advent of the Industrial revolution, black smoke from chimneys started to pollute the atmosphere. Some of this smoke and soot made up of small particles of unburned coal deposited itself everywhere including trunks of trees. This blackening of the trees in Industrial areas such as Lancashire had a dramatic effect on the survival of the silvery peppered moth. The moth that for thousands of years developed its silvery colour, because it could not be detected against the silvery trunks and stems of trees, was now becoming visible from a long distance by its predators. It simply lost the advantage of camouflage. The contrast in its colour against that of the darkened tree trunks made it a simple catch by its predators. The silvery peppered moth simply disappeared from these areas. In its place however a black mutant of the species increased in numbers. This black moth soon started dominating these areas.

Even though these events were taking place at the time that Darwin was looking for evidence to get full acceptance of his Theory, he was not aware of them. Such an example would have provided for him exactly the evidence he needed.

There is a sequel to this example. Years later, when the amount of soot in the atmosphere was reduced, the tree trunks returned to their original colour and the original silvery peppered moth reappeared, whilst its black mutant was greatly reduced in numbers.

It is important however in looking at this example to appreciate that randomness doesn't come into it. The peppered moth simply comes in two different colours black and silver. One is a mutant of the other.

-.-.-.-.-.-.-.-.-.-

Darwin assumed in his theory that many more of a species are born than can survive to adulthood. Furthermore he assumed that many more are born than can survive in a particular environment due to limitations of food, the threat from predators, etc. Let us consider an example where the large number of offsprings has the chance of following evolution through natural selection.

In order to breed, the Atlantic salmon returns from salty water where they normally live to fresh water in exactly the same place where their ancestors bred for many generations before them. They do this, year after year, and from generation to generation. It is believed that the salmon find their way back to their ancestral home through the sense of smell. The spawning fresh water stream is normally cold, clear, fast flowing and with gravel and rocks at the bottom. To get to it, salmon swim upstream at the rate of 6 to 7 km per day. They jump over obstacles they find in their path, against the water flow, over waterfalls, sometimes in excess of 3 metres in height. During this migration period as well as the period until they mate, they eat nothing. As soon as they arrive at their spawning ground their colour changes from a dull grey to some bright hues. The male develops a hooked lower jaw and a humped back, whilst the female digs out a nest for the eggs. The males fight between them in some ferocious battles using their newly acquired hooked lower jaws, for the right to fertilise the eggs which the female deposits amongst the gravel. After fertilisation, the female stirs up the bottom of the river to cover the eggs and protect them. Male and female then return back to the sea. With some species they do not return back to the sea. As soon as their breeding function is completed, they simply die. The young hatch within about two months and they then swim to the sea.

This very elaborate breeding procedure must have evolved over thousands or millions of generations. Salmon have probably realised that the chances of their eggs surviving are higher in the river than in the sea. That is why they are following this strange breeding course every year. Salmon lay around 20,000 eggs. The chances of evolving through natural selection, according to Darwin's theory must be quite high. Of the 20,000 eggs only very few reach sexual maturity.

This is completely different to, say, the case of man where usually only one offspring is born, and that offspring is looked after very carefully. Thus the chances of evolution due to natural selection are non existent.

Surely one can easily see that the case of salmon with its 20,000 eggs, the random nature of the hatching of the eggs, the large number of predators in the river, presents a completely different case to say that of man with normally only one offspring that is given the best possible care for many years.

Thus Darwin's assertion that "many more of a species are born than can survive to adulthood" and that "many more are born than can survive in a particular environment due to limitations of food, the threat from predators, etc" is fully vindicated for the particular example of the spawning of the salmon.

Another example where numbers are important is the conifer cone. One conifer cone can produce several million grains of pollen for fertilisation. An average tree can produce a few thousand cones. The chances of a grain of pollen finding the style of another individual of the same species and causing fertilisation, which then results in the formation of a new tree, are pretty remote. This is another obvious example where a selection procedure provides the possibility to select the best, through very large numbers. Obviously the selection criteria are very difficult to define in this case. There must be a large degree of randomness as to which grains of pollen actually succeed.

Another good example is that of an oak tree which can produce up to 100,000 acorns in one year. If only one acorn out of the 100,000 were to become a growing tree every year, the population of oak trees in the long run would multiply enormously, as oak trees can live over 100 years. I have estimated (using a computer program) that if one acorn from each tree were to become a growing tree, the whole surface of the earth will be covered with oak trees in less than two hundred years. Please see Appendix 2 for the tabulations. Yet we know that the earth's surface is not covered with oak trees. Therefore the success rate of acorns resulting to trees is lees than 1 in 100,000. Obviously the degree of randomness once again is huge. This is exactly what Darwinian Natural Selection is all about.

So one could say that Darwinian Natural Selection is fine the way Darwin described it, i.e. where the number born in a species is much greater than those that survive to sexual maturity.

As we shall see later in this book, there are many cases, such as the case of man (where today's survival rate in most countries is around 97%), which do not always conform to Darwinian Natural Selection.

-.-.-.-.-.-.-.-.-.-.-

The examples I am presenting in this book cover a wide range of insects, plants, fish, reptiles, and birds as well as mammals including man.

I will start with an example concerning ants. I carried out the following investigation myself. I took a tall, clean drinking glass deep into the Vietnamese jungle. I put some bread and sugar into it. I attracted the attention of various species of ants around small areas by dropping some bread and sugar on the ground in different locations in the jungle. I then placed the glass close to the bait on the floor. Ants were attracted to the bait but none of them could climb up the polished glass. I then took a few ants from 3 different species and put them successively in the glass. None of them could climb out.

I then put the same glass with the same bait in different locations within or close to homes not very far from the jungle. The domesticated ants had no problem whatsoever in climbing up the polished glass and carrying away the bread within it. I tried this with six different species of domesticated ants. None of them had any difficulty whatsoever in climbing up and down the glass.

The obvious conclusion is that the domesticated ants have managed to develop a means by which they can climb up polished and slippery walls in order to get to various foods in the house. One possibility which should be investigated is that these ants may have developed suitable sucking pads on their claws that allow them to crawl vertically on polished surfaces. Evidently such a development has taken place during the last few thousand years when such surfaces appeared in the homes of humans.

I present this as a clear example of rapid evolution – evolution that took place under pressure for survival. It is highly unlikely that the Darwinian (or rather neo-Darwinian) explanation of random mutations can fully explain these observations.

During my exchange of correspondence with various scientists around the world on my findings on ants, it was brought to my attention that the American gecko has exactly the same characteristics. There are two native species of gecko in the United States, the leaf-fingered gecko (*Phyllodactylus tuberculatus*) of extreme S California and Baja California, and the banded, or ground, gecko (*Coleonyx variegatus*) of the deserts of the SW United States and N Mexico.

The first is a house dwelling gecko and it has characteristic pads on the underside of its feet which gives it the ability to climb up on smooth surfaces as well as to cling upside down on ceilings. The second lacks the characteristic foot pads of the domestic species.

The reaction of some neo-Darwinians to whom I brought this example to their attention was that the domestic species already had the pads before it became domesticated. I should say that four neo-Darwinians that made this comment made it with a high degree of exuberance and in fact one of them even with a touch of irony. This of course raises some serious questions:

Where are those geckos that used to live outside homes and had pads on their feet? Is it fair to assume that absolutely every single one of them moved into homes? Why only the geckos with pads moved into homes and those without pads stayed outside homes?

The pads contain microscopic backward-projecting hairs covered by tiny pads that generate an adhesive force through van der Waals attractions (Inter-molecular forces).

The gecko example has now given me confidence to mention some personal experience whilst I was staying in a hotel in Vietnam. I had some biscuits in a shopping bag and suspecting that there might be mice in the hotel I hanged this bag on a stainless steel hanger more than one metre from the floor. This highly polished stainless steel hanger was about 30 centimetres from the nearest wall and about 50 centimetres from a wooden wardrobe. During the night I realised that a mouse had entered this shopping bag with the biscuits. I do not know how it got there. The easiest explanation was that it went to the top of the wardrobe and then jumped into the bag that might have had a slight opening near the top. Or it went up the wall and then jumped into it. This mouse must however been some jumper to achieve this feat. But I have my suspicion that this mouse may have climbed up one metre of the polished stainless steel hanger. To achieve this feat this mouse must have possessed pads on its feet.

-.-.-.-.-.-.-.-.-.-.-

Let us now look at an example concerning the human race, which has baffled me for a long time.

The disease of malaria is mainly found in the tropical and subtropical regions of the world. It is a disease spread by various mosquitoes and which throughout the centuries has wiped out complete populations. It is generally regarded as one of the most serious diseases ever, and even today over one million people die of malaria every year. However two different types of mutations in two different parts of the world have created cells that induce immunity to malaria.

The first type of mutation appeared in the region of Central West Africa. The acquired mutation changes the shape of the blood cell to the shape of a sickle, when the cell becomes acidic. When a malaria parasite enters the red blood cells its waste products make the cell acidic. As the deformed sickle shaped cells pass through the blood stream, the body's defensive system targets these modified shape cells for destruction. In destroying these cells however, the invading parasite attached to it, is also destroyed.

Millions of people have been saved through this mutation, which is very widespread in various countries in West Africa. This mutation has however introduced a new problem to some of the people that have acquired these modified genes. If the parents of a child are both carriers of this modified gene, then according to the theory of genetics, there is a high possibility that one of four children will suffer from an acute form of anaemia called sickle cell anaemia. This anaemia can produce blood clotting which can be very painful. It also leads to frequent premature deaths. If only one of the parents is a carrier of the modified gene then the children can be either carriers or not carriers, but none of them will carry the full disease, which can be lethal.

In India, the Middle East and some Mediterranean countries where malaria was also very widespread over the last 4,000 years, a different type of mutation created a similar immunity to malaria. Even though millions of lives were saved from malaria through this mutation, a new anaemia was created called Thalassaemia. Similar to sickle cell anaemia, there is a high likelihood that Thalassaemia can be lethal to one child in four, if both parents carry the modified genes.

The obvious questions that arise are "Why did these two beneficial types of mutations take place in the areas of the world where they were so desperately needed? Why did they not occur in Northern Europe or China for example or indeed in other countries where malaria is not so common?"

The issue of a beneficial mutation occurring at precisely where and when needed becomes even more apparent when one considers the fact that there are hundreds of different such mutations concerning theses two diseases. Recent studies for example have identified at least 347 mutations linked to the beta Thalassaemia trait. It appears as if some force has been working extremely hard to ensure that such mutations take place to ensure the survival of human beings in the Malaria afflicted regions of the world.

Don't you think it is rather strange that 1 in 20 Indians and 1 in 7 Greek Cypriots is a carrier of one of these mutations whilst such mutations are virtually unknown in Scandinavia?

One could argue that the reason that the Thalassaemia mutation is so apparent in the regions of the world where Malaria was so widespread is that there are now many carriers as confirmation of these mutations. If there was such a mutation say in Sweden that mutation would not become apparent and the person concerned would not see any benefits from it. If there were any in Sweden and they intermarried, then we would have known. Otherwise we may never find out, because a carrier does not necessarily suffer from any obvious ailment. However it is a fallacy to be looking at the number of people that survived and their percentages relative to the rest of the population in the region. The fact is that there have been at least 347 mutations in that region. And none in Sweden – as far as we know

The Darwinian answer that these mutations happened by sheer chance cannot be contemplated. This is too much of a coincidence! A more positive answer is needed to explain these two unrelated types of mutations that saved so many millions of lives. This is what I referred to as a problem that needs to be addressed. If Darwinian Theory cannot provide a scientific answer then obviously Darwinian Theory has to be modified. This is precisely what I am trying to do through my Hypothesis.

-.-.-.-.-.-.-.-.-.-

When considering the total number of mutations for a species during a certain period, the main factors of influence are the numbers of creatures of the particular species plus their life cycle and the number of offspring. For example for humans today there are 7 billion people alive that normally reproduce at between 20 and 30 years with an average of about 3 children per couple. The rate of birth in the World today is around 100 million people per year. This is equivalent to about 12,000 births per hour.

Rabbits and hares with a life span of about 10 years reach sexual maturity in about 6 months and produce up to 8 litters per year with about 5 to 8 young per litter. In places like Australia there are many millions of rabbits. I have used a computer program to evaluate the number of rabbits born per year and to summate the number over a period of time, starting with one pair of rabbits. Assuming two female rabbits per litter and 4 litters per year I found that the total theoretical number of rabbits within 8 years would be over 34 million.

Please see Appendix 3 for details of this calculation.

It is therefore obvious that the possibility of random mutations for rabbits is much higher in rabbits than for humans. After all, their DNA code is very similar to that of humans.

For many examples that follow, I refer to the number of generations that are relevant for a particular characteristic to evolve. This is based on the total number of years for the period under consideration and the estimated reproductive life span of the species concerned.

It is generally accepted that Homo erectus who lived about one million years ago is the direct predecessor of the human race. This is the point that the human race started its direct evolutionary development. If we were to assume that the average age of a generation of these creatures was around 10 years (assuming that they reached sexual maturity at that age), then over a period of one million years, this is equivalent to 100,000 generations. Thus all the changes from the primeval beings to Modern Man cannot be attributed to more random mutations than can be attributed to the 100,000 generations that led from a primeval creature to modern man.

Rabbits would have gone through many millions of generations during this period of one million years. Also there were many more rabbits alive during this period than there were creatures in the direct line of ascend of man. The corresponding number of mutations would therefore be infinitely more amongst rabbits than with the predecessors of humans. Yet there is no evidence of any significant genetic changes to any of the characteristics of the rabbit.

I know that Darwinians skip over this issue as it does not concern them but it is a very serious matter that needs to be addressed and discussed. We have on one side just 100,000 generations with very few individuals and on the other side we have many millions of generations (over the same period of time) and many millions of individuals. Yet we have had such mutations that changed a beast to modern man in the first case and we have had no mutations that made any changes to the rabbits whatsoever in the second case.

It is unscientific to be dogmatic and skip such issues. Even Socrates and Galileo were given more opportunity to present their point of view than Darwinian dogmatists give to revisionists.

It is estimated that on average there is a new mutation per birth of a human. But it does not necessarily mean that such mutations are inherited by the next generation as added characteristics that will modify the human race. Some of them may of course do so. Otherwise how can one explain the fact that the average American today is some 10-12 centimetres taller than his ancestors of only 100 years ago?

All I am trying to emphasize here is that I cannot see how the large number of characteristics acquired by modern man, such as his powerful over-endowed brain, could have been acquired in just 100,000 generations through Darwinian Evolution.

As I stress elsewhere in this book, throughout the period from Homo erectus to Modern Man there are some periods during which these beings virtually disappeared. The main period is that from about 500,000 years ago to about 110,000 years ago. Very few fossils have been found that concern this period. Also it is estimated that towards the end of the Neanderthal period there were only about 2,000 Neanderthals before Cro-Magnon appeared.

Following these arguments, questions arise as to whether human evolution is not based entirely on randomness but possibly on some other aspect of human genetics. For example, is it possible that two (or more) processes are simultaneously contributing towards evolution?

Or is it possible that evolution works in such a way that changes are occurring continuously sometimes producing dramatic effects in a short period of time and sometimes producing small changes that are not always easily observable? After all, as I said above, there must be an explanation as to why the average American man is some 10-12 cm taller than his ancestor of 100 years ago! Furthermore this dramatic change in the average height of the American man (and of men of other prosperous nations) did not occur suddenly, but gradually over the period of the last hundred years. If this happened by random mutations why did it involve the men of the most prosperous nations, and not the men of the deprived nations? These are important issues that cannot be pushed to one side!

Obviously I am not rejecting the fact that for a genetically acquired characteristic a change in the DNA is essential. That goes without saying. After all, the science of genetics has now reached such levels that this is a matter that nobody can dispute.

What I am disputing is the way that the mutations are created within the sex cells. I am simply saying that in some cases random mutations cannot provide acceptable explanations.

Perhaps I should stress here that I am prepared to possibly contemplate (rather than unconditionally accept) Darwin's explanation of the giraffe's neck, as the giraffe has had tens of millions of years to develop such a long neck through random processes. Tens of millions of years is quite possibly within the range where Darwinian evolution could apply for an animal such as the giraffe.

-.-.-.-.-.-.-.-.-.-

Some 65 million years ago the dinosaurs that were established as a major group of animals on earth for over 160 million years, suddenly became extinct. It is thought that the reason for this was a large asteroid that collided with earth at that time. This collision resulted in a big cloud of dust that virtually surrounded the whole of the earth.

The climatic conditions changed, due to this dust cloud, and the dinosaurs could not change fast enough to be able to survive in the new climatic conditions on earth. It is quite possible that numerous other species that could not change their characteristics fast enough also perished due to those sudden and drastic climatic changes.

Other animals possibly changed very quickly and managed to survive.

This change under pressure is an excellent example where random changes could have offered few chances of producing the right characteristics in the time interval that these changes needed to take place to ensure their survival in the new environment.

-.-.-.-.-.-.-.-.-.-.-

One other aspect that I have not introduced in the previous arguments, but which is of great significance, is the matter of survival of the fittest. As I discuss in other parts of this book, many species bring many thousands of offspring to this world. Most of them do not reach the age of sexual maturity. In the case of salmon for example the female lays 20,000 eggs. Only a few reach the age of reproduction.

Even in the case of very caring parents, such as for example the elephants or the African wild dogs, the chances of their offspring surviving to sexual maturity are in most instances below 50%. For most animals in the jungle, the chances are much below that. This low ratio of survival offers nature the chance of ensuring that only the fittest survive. The jungle is not a place for weaklings.

The important concept that defines the successful outcome in evolution is the ratio of those that survive compared to the total born.

It is apparent from this, that if in some way all that are born survive to the age of sexual maturity, then there will be no selection process and the matter of survival of the fittest will not arise. This is obviously what exactly happens in the case of man. Every single child born is looked after in such a way that it will be ridiculous to even contemplate using the term of "survival of the fittest".

However one could look back and comment that not so long ago diseases such as malaria, tuberculosis, cholera and so on were responsible for the deaths of millions. This is true of course and in the previous section I looked into this particular aspect with emphasis on malaria. In spite of these figures however the actual percentage of children surviving to sexual maturity was quite high. There is absolutely no comparison between the figures for the rate of survival of animals in the jungle even of the wild dog and that of humans, even if we go back a few thousand years. Man looked after his offspring in a special way that no other creature could or did.

The percentage of children in the western world of today that reach sexual maturity is around 97%. This basically eliminates the element of survival of the fittest when applied to man in the western world.

It is true of course that this figure is much lower in some countries in the world today, mainly due to famine and disease. It is difficult however to visualise a situation where in a country that starvation is a major problem, the principle of the survival of the fittest leads to a new species of man that can survive on very little food.

Thus the principle of the survival of the fittest may apply to all other creatures, even though only to a lesser extent for some of them. But the principle of the survival of the fittest definitely does not apply to man.

As Darwin said, evolution is fine the way he described it, i.e. where the number born in a species is much greater than those that survive to sexual maturity.

-.-.-.-.-.-.-.-.-.-

I would like to mention here another example concerning the human race.

After a man climaxes, during sexual intercourse, there is a definite time lapse before ejaculation takes place. This small time interval is not normally noticeable as it is only one to three seconds. This is probably a unique feature that no other animal possesses. This period between climax and ejaculation is of great importance for people who want to avoid conception, as it offers men the opportunity to withdraw before ejaculation. In fact through the years, this has been the most commonly used method of contraception by men throughout the world.

This time delay between climax and ejaculation varies amongst people. It is obvious of course that those men with a long time delay between climax and ejaculation are much more successful in preventing a possible pregnancy, whilst those with a short delay will have problems in avoiding a pregnancy.

Thus men with a short time delay will produce more children than those with a long delay. This is a typical example of natural selection of the Darwinian evolution whereby one would have expected the men with the short time delay to soon dominate in numbers around the world. In fact theoretically, this time should have continuously been diminishing, to the extent of tending to be close to zero.

However we know very well, that this time delay is not zero, but a definite period, which provides the facility of contraception. The question then arises: Why has natural selection not reduced this time delay to zero?

There are of course those that will say that this time delay is not a genetically acquired characteristic. But if it was not a genetically acquired characteristic, then it would not have been present in nearly all men. It is as much a genetically acquired characteristic just as much as the fact that all men have two legs, two feet and one head.

-.-.-.-.-.-.-.-.-.-

Let us consider a few more examples.

Whenever we acquire an injury, irritation, infection or inflammation of the skin, white blood corpuscles are carried by the defensive systems of our body to that location to fight the infection.

The white corpuscles can be of great importance in this instance. However one has to consider the deeper significance of this genetically inherited function of the body. Is this a function that is of paramount importance to survival? Basically one has to consider what could have happened to our ancestors during the last 100,000 or so generations, before they acquired this characteristic of the white blood corpuscles rushing to the defence of the body, whenever they suffered an injury on the skin.

According to Darwinian Evolution, someone, by sheer chance or coincidence acquired this characteristic. This individual obviously had a higher chance of surviving if he or she had an injury. If, this acquired characteristic was dominant and inherited by the descendants of this individual, then obviously these descendants had a higher chance of survival compared to the other individuals that had not acquired this characteristic. However one has to postulate the time that it would require for such a characteristic to become universal, bearing in mind that the remaining of the "human race" would not have become extinct overnight, through a small injury to their hands. After all they managed to survive for probably hundreds of thousands of years (i.e. tens of thousands of generations) without this advantageous characteristic.

-.-.-.-.-.-.-.-.-.-

It is normal to describe differences amongst people of various races as being the outcome of Natural Selection or survival of the fittest.

Thus for example, the Eskimos are supposed to have acquired their ability to live in the cold, through the fact that most of the people that could not adapt to those conditions, simply died before reaching the age of producing any offspring. Those that were able to live in the cold environment of the Eskimos, reproduced enough offspring to ensure the continued existence of the Eskimos. Of course one could assume that being sensible human beings, if they felt that they could not survive in that cold environment, they simply would have walked away to warmer areas. So, those that stayed behind, were those that could withstand the cold weather. Eventually all Eskimos have acquired, the characteristics required to survive in those cold parts of the world.

Such characteristics include shovel-shaped incisor teeth which have evolved due to the need to chew the raw or virtually raw meat which is their basic daily diet. They also possess an exceedingly narrow nose which reduces the effects of the cold environment in which they live. They are normally short and squat which is the best shape for a human body not to lose heat and to keep warm. They have developed a layer of fat under their skin to be able to withstand the very cold weather.

Bearing in mind that the Eskimo's History can extend to a maximum of 4,000 years, and that the total population has been and is around 50,000 people, it is obvious that it would have been impossible for randomness to have created such essential traits within such a short period of time

-.-.-.-.-.-.-.-.-.-

Let us now look at individual acquired characteristics of people.

Let us for example consider the shape of the human eye. The easily recognized shape of the Mongolian eye is supposed to have developed through Natural Selection to be able to withstand the very cold and strong winds, one associates with that part of the world.

So, according to Darwinian Natural Selection, there were at one time people with Mongolian shaped eyes and people with ordinary shaped eyes. Those with ordinary shaped eyes that happened to find themselves in Mongolia, could not survive and either died before producing any offspring (or did not produce enough offspring before passing away) or simply moved away from Mongolia. So, eventually according to Darwinian Natural Selection, the Mongolian shaped eye has dominated in Mongolia.

One important question does however arise:

Is the Mongolian shaped eye a necessary characteristic to survive in Mongolia or simply a desirable characteristic that provides more comfort? It would take a bold person indeed to suggest that the shape of the Mongolian shaped eye is an absolute necessity for survival in Mongolia. The only logical conclusion is that the Mongolian shaped eye has evolved through a means that is hardly related to Darwinian Natural Selection. It was simply a localized requirement that developed because of the local weather conditions.

After all, Darwinian Natural Selection is strictly "a consequence for the struggle for existence", or "survival of the fittest", not of creating more comfort to individuals.

Incidentally there are those who believe that the shape of the eyes of the Eskimos has specially evolved in such a way as to reduce the effects of the very high reflection of sunlight off the snow.

-.-.-.-.-.-.-.-.-.-

Continuing with the human eye, it is well known that when we try to look sideways to the right or left, there is a certain angle at which we simply cannot see anything. This "blind spot" has been the cause of many traffic accidents because as drivers come out of a side street into a main road cannot sometimes see the traffic coming along the main road. The reason for this "blind spot" is that the light rays entering the eye get focused onto the retina at the back of the eyeball. This focusing of the rays enables us to see and interpret the view in front of us. However this information has to be passed onto the brain. This is done though a large number of fibres based behind the retina, where each fibre corresponds to light receptors on the retina. However in order to come out of the retina these fibres are bunched together and get out of the eyeball through a certain part of the retina. This escape area of the fibres cannot obviously have any light receptors on top of it. Thus light rays focused on this part of the eye are simply not detected.

If the eye had followed the easiest form of evolution it would have followed a route of symmetry. For example most flowers are circular in shape with all the sepals and petals emerging symmetrically outwards along a radius from the centre of the flower. Virtually all stems of plants are also circular. The iris of the human eye is circular as well in order to produce the focusing like a virtually perfect lens.

So why is this bunching of the fibres not placed symmetrically in the centre of the retina? The answer lies in the fact that if it was placed in the centre of the retina viz. on an axis from the centre of the iris, then most of the light rays that come from objects directly in front of us would not be detected by the retina receptors as there would not be any receptors at this critical area of the retina. So the bunching of these fibres has been placed in the best possible position to cause the minimum amount of intrusion.

Can you see this being produced by chance?

In Chapter 6 of his "Origin of Species" Darwin clearly says:

"As natural selection acts by life and death I have sometimes felt great difficulty in understanding the origin or formation of parts of little importance."

What a wise statement!! I wish some more modern Scientists were to take notice of such an astute and sensible statement.

-.-.-.-.-.-.-.-.-.-.-

Let us now look at some further examples of acquired individual characteristics that can hardly be explained through Darwin's Natural Selection as a consequence for the struggle for existence.

It is generally accepted that the Negro race, one of the three main anthropological divisions of mankind, appeared in the African region south of the Sahara, some ten thousand years ago.

The main characteristic of Negroes is the dark colour of their skin. This is due to the high concentration of melanin in their skin, which increases the rate of absorption of the useful component of the sun's radiation, which in turn provides additional protection to the body. Equally important is the fact that the black body reflects the harmful Ultra Violet radiation. These rays can cause cancer to the skin of the human body. This is evidently a desirable characteristic, but obviously not a necessary characteristic, as millions of white people that now live in the same regions of Africa can confirm to us.

As this characteristic is not absolutely necessary for survival, then the Darwinian Natural Selection as a consequence for the struggle for existence cannot surely be applied in this case. According to Darwinian Theory one would have to assume that a random mutation some ten thousand years ago led to the people of that region of Africa to produce a higher concentration of melanin in their skins.

The main question arises: What are the chances of this random beneficial mutation occurring in that region of the World as opposed to the other regions of the World?

Of course the answer is that the chances are reasonably high, as the chances have to be equally distributed amongst the people alive at that time, viz. ten thousand years ago. Let us take a figure of say 10% (assuming that 10% of the people alive at that time lived in that region of Africa where Negroes suddenly appeared).

Let us now look at further features of Negroes that must have evolved at that time or around that time, i.e. ten thousand years ago.

The body of the Negroes has a high concentration of sweat glands. This produces a higher amount of sweat that on evaporation cools the skin of the body, which in turn ensures a degree of comfort in the very high ambient temperatures.

Like the colour of the skin of Negroes, this is a desirable characteristic, but obviously not a necessary characteristic, as millions of white people that now live in the same regions of Africa can confirm.

Yet again a very important question arises: What are the chances of this random beneficial mutation occurring in that region of the World as opposed to the other regions of the World?

For the second time let us take the figure of 10% as appropriate for the chances of this characteristic appearing in Africa as opposed to any other part of the World. This can obviously be justified through the assumption that at that time 10% of the world's population lived in that region of Africa.

For those of you whose mathematical knowledge is slightly higher than average, I am sure that you can immediately evaluate that the chances of these two characteristics appearing together (not necessarily simultaneously) in Africa are getting lower. In fact the chances are 10% of 10%, i.e. the chances are now only 1%.

Let us now look at a further acquired characteristic of Negroes. The hair of Negroes is extremely curly. In fact their curliness is virtually unique amongst the rest of the human race. The curliness of the hair provides additional surface area on top of the head, which helps to increase the evaporation of sweat on top of the head, which again provides stability of the body temperature.

Like the colour of the skin of Negroes and their higher number of sweat glands, this is a desirable characteristic, but obviously not a necessary characteristic, as millions of white people that now live in the same regions of Africa can confirm.

Once more a very important question arises: What are the chances of this random beneficial mutation occurring in that region of the World as opposed to the other regions of the World?

For the third time let us take the figure of 10% as appropriate for the chances of this characteristic appearing in Africa as opposed to any other part of the World.

The chances now of these three inherited characteristics occurring in Africa as opposed to any other part of the World, are getting very low indeed. In fact if we were to apply the previous assumptions I am sure that most of you would agree that the chances are one in a thousand.

Obviously as we consider further characteristics of the Negroes we realize that the chances of these events happening randomly in Africa are getting infinitesimally small.

Yet as we can all testify, these characteristics do exist amongst the Negroes and as we can all testify these characteristics are inherited. In other words these characteristics form part of their genes.

However, how many of us are willing to bet our bottom dollar that all these characteristics are the result of random mutations that occurred in successive generations during the last 10,000 years or say 1000 generations?

One corollary to the above is the question as to why these mutations did not take place in other places such as Northern Europe, Russia, China, etc. where the sun's radiation does not present a particular problem. After all if one could assume that 90% of the world's population lived outside the sub-Saharan region, then the chances of any one of such mutations occurring in these regions are nine times higher.

One other characteristic of the Negroes relative to Northern Europeans is the colour of their hair which is normally black due to the very high content of melanin as a result of the strong sunlight.

The basic reason that the hair of many Northern Europeans is blond is because of the very low content of melanin in their hair due to the absence of continuous strong sunlight throughout the year in Northern Europe. Melanin is a pigment formed by the oxidation of tyrosine which is a colourless amino acid. Melanin is contained within minute granules and the colour depends on the number of granules as well as the density of the pigment within the granules. Blonds simply have a low number of granules and a low density of melanin in the granules of their hair.

For some strange reason this chemical characteristic has become a genetic characteristic for both Negroes and blonds. This deserves further investigation. This might even lead to a third option for Evolution. In this book I describe Dual Evolution but maybe this is only one additional option to Ordinary Darwinian Evolution. There may be others.

-.-.-.-.-.-.-.-.-.-.-

If we compare the general external features of an African Negro with those of an Australian Aborigine we would immediately notice some obvious similarities. The colour of their skin is the most striking similarity. However the hair of the Aborigines is not curly as that of the Negroes. Further study leads to the obvious conclusion that these two groups evolved separately. But why is it that the colour of their skin is so similar? Bearing in mind that these two groups evolved in very similar climatic conditions and that as discussed above, the black colour of the skin is a much desired feature for these climatic conditions, I find it insulting for someone to insist that these two features evolved independently through randomness.

We could also look at other people who live close to the Equator around the World and observe such similarities. Thus many South and Latin American people as well as some in the South and South East Asia do have these similarities albeit to a lesser extent. We should of course bear in mind that there has been a big influx of people from other regions in some of these areas. There are however some notable exceptions. For example even though the people of Sub Saharan Africa around the 10th to the 20th

Parallel are virtually all Negroes, these characteristics are not so obvious in India and in the case of Malaysia, Cambodia, and Vietnam they are completely different.

The colour of their hair however is frequently striking. Frequently the shape of the hair is also worth noting. For example any visitor to Vietnam cannot but notice that virtually all girls have very long, very straight, shiny, smooth, black hair. Looking at the back of their heads they almost all look identical.

-.-.-.-.-.-.-.-.-.-

One of the most important statements frequently made by Darwinians is that nothing that is acquired in a generation can be inherited.

However, as mentioned elsewhere in this book certain viruses that can cause cancer in animals can transfer information from RNA to DNA using an enzyme with the descriptive name "reverse transcriptase".

Also in a very important experiment carried out in the early seventies, scientists trained some planarian flatworms to respond to light when placed in darkness. They then chopped them and mashed them up before feeding them to some other flatworms.

The results were astounding. These flatworms could also detect light. So the modified DNA in the memory cells of the first group of flatworms was passed on to the memory cells of the second group of flatworms.

The most astonishing result however was that when these flatworms were allowed to breed, their offspring could also detect light. So features acquired in a generation were genetically transferred. The DNA was permanently modified.

This experiment provides justification to the scepticism of many groups of people who object to genetic manipulation of any kind. They argue that at this present juncture we lack the scientific knowledge to know the exact consequences of even the most apparently innocent genetic manipulation of food. We are working in the dark. Unfortunately the pressures of commercial advantage are forcing many scientists to carry on relentlessly with their work in genetic manipulation.

Perhaps some of these people should be reminded of the case of Marie Curie who soon after inventing radioactivity started using it for the benefit of mankind. But as she was not aware of the hidden dangers of handling such dangerous substances, she soon became the victim of radioactivity herself. She died of cancer.

As the famous German philosopher Goethe said "There is nothing worse than ignorance in action".

Let us hope that common sense will prevail before it is too late.

-.-.-.-.-.-.-.-.-.-

CHAPTER 3

A NEW HYPOTHESIS

Having established the need for an alternative explanation to Darwin's Theory of evolution for some unexplained phenomena, let us look at the following proposal:

Consider the case of the first hominid, Australopithecus Africanus who lived from about 3 million years up to about two million years ago. He was probably the first "hominid" to walk on two legs. Walking on two legs, allowed him to start to use his hands for other things such as making tools. He was about 1.2 metres tall and weighed about 40 kg. He was not as fast as some of his predators so he had to invent suitable but crude weapons to fight these predators.

To create tools however he had to have had the mental capacity to do so. Just because he started walking on two feet, which in itself may have been a progress compared to his predecessors, he did not necessarily become more advanced in any other capacity. His brain was only about 440cc, which is only about 10% greater than that of the chimpanzee. This did not make him into a genius. But he had to come up with an answer quickly otherwise his predators would have "made a meal" of him.

His first weapons were probably swinging branches of trees much as some apes do today. His slightly more advanced brain enabled him to create some crude weapons.

It appears that the two developed together, i.e. the progress in his tool making ability and the increase in the size of his brain.

It took about one million years for his brain size to increase by 50% to about 660cc. This new creature with the bigger brain that lived from about 2.2 million years ago to about 1.5 million years ago is called Homo Habilis (which means handy man). He created tools and weapons that were much more advanced than those of his predecessors. Obviously any tools or weapons that he developed were still crude and only served very elementary purposes.

Going back to our previous analogy and considering an average age per generation of 5 years for Homo Habilis, we get a total of 200,000 generations during which the brain capacity increased by 50% and his tool making ability improved by a large factor. His appearance changed somewhat, his body increased in size and his molar and premolar tooth size reduced.

But very little else improved during one million years and 200,000 generations of Darwinian evolution.

Exit Homo Habilis, enter Homo Erectus. Yes, the next creature to appear about 1.5 million years ago was Homo Erectus with Homo Habilis vanishing from the scene. He was the first recognizable human creature. He was the first real hunter of big game using better-made weapons. To achieve this, they must have been hunting in groups. This has led some archaeologists and anthropologists to the conclusion that Homo erectus developed some form of speech. However if that is the case his speech did not have to be very elaborate, probably consisting just of various gradations of howling and grunting. After all, group hunting in animals is not uncommon. For example the African hunting dogs always hunt in packs and a big pack can even bring down a lion by using ingenious group tactics.

The most important aspect of Homo erectus was the size of his skull, which grew to about 1000cc. He also started using fire towards the end of his span, which happened about 500,000 years ago. He probably used fire from burning trees that caught fire due to a lightning striking them.

So he achieved all this in about one million years or about 200,000 generations.

-.-.-.-.-.-.-.-.-.-

For some unknown reason, around 500,000 years ago Homo erectus started moving out of Africa and into more temperate zones in Europe, the Middle East and Asia. But then suddenly there seems to be a problem. We have lost their trace. There seems very little evidence of fossils for nearly 400,000 years. Very few skulls have been found to account for these 400,000 years. There was one found near Heidelberg, which was dated about 400,000 years old, one near Steinheim and one in Peking which were dated about 300,000 years old and one near Swanscombe dated about 250,000 years ago.

The size of these skulls of the Heidelberg, Steinheim and Swanscombe man range from about 1100cc to about 1250cc.

Some people have called this period viz. from about 500,000 to around 110,000 years ago as the "missing link". Because of the finds mentioned above some people call it the missing link of 300,000 years and some the missing link of 200,000 years. The numbers are not that important. What is important is that there is a missing link of a few hundred thousand years.

There are other periods much earlier than this period where there seems to be a missing link but this period is regarded as the major unexplained "missing link".

I find it amazing that so many people disregard this issue of the missing link as if it was a minor problem. For surely, if there were not many people around there could have hardly been much evolution taking place. The absence of fossils can only mean the absence of living creatures. Disregarding this issue as a minor one cannot lead us to any sensible conclusions. It has to be treated as a major problem that has to be addressed and discussed openly and extensively.

Not many people have tried to give any explanations as to the disappearance of Homo erectus mainly because many Scientists disregarded it as if it is a period of minor significance or as a period that never existed. One could postulate a collision with an asteroid which affected Homo erectus drastically (just like 65 million years ago another asteroid resulted in the disappearance of the dinosaurs). Or one could postulate a disease such as malaria which virtually wiped Homo erectus off the surface of the earth. I believe that the explanation that I propose further down makes more sense.

Suddenly for no apparent reason around 110,000 years ago there seems to be an abundance of fossils of a new creature named Neanderthal man, as one of the first finds was in the Neander Valley in Germany. The Neanderthal, who lived from about 110,000 years ago to about 35,000 years ago, was much bulkier than any of his predecessors and had a huge skull that varied from about 1400cc to about 1600cc.

The distinctive features of the Neanderthals were a low, sloping forehead, large brow ridge, and a long face with a dominant projecting nose. They were stocky, sturdy people with features that helped them survive in the very cold climate that existed in Europe at the time. Their tools included flint scrapers for skinning animals to make clothes. They also buried their dead. Some anthropologists believe that they could also speak. Again if they spoke, their speech may have been very limited, not that very different from his ancestors who may have used various gradations of howling and grunting.

But then suddenly within a period of just a few thousand years Cro-Magnon, the Modern Man appears and the Neanderthal disappears. This happened around 35,000 years ago.

The main questions arising from the above are:

1. Why are the fossils between 500,000 years ago and 110,000 years ago so scarce?
2. How and from where did the Neanderthal man appear suddenly?
3. Why did the Neanderthal disappear with his place being taken over by Cro-Magnon again within a very short period of time?

Let us look at the first question in relation to the size of the head and the continuous pressure to have a bigger brain. The size of the skull of the Homo erectus at around 500,000 years ago was about 1000cc and at that stage continued to increase in size. As indicated by the few finds of skull during the "missing link" period, the skull increased to about 1250cc. But the height of the Homo erectus was more or less the same for hundreds of thousands of years at an average of about 1.55m.

So we have two issues concerning that period of time. Firstly the size of the skull was getting bigger and secondly the number of people started to become smaller. Were these two issues related?

If these two issues are related then this can lead to one possible explanation. The bigger size skull was creating birth problems because of the size of the female birth canal. In other words many females died whilst giving birth.

That explains the "missing link". Few individuals survived, that is why we have only very few fossils.

That brings us to the second question as to how Neanderthal man appeared suddenly – apparently out of nowhere and with very few predecessors.

The answer lies in the fact that somehow Neanderthal achieved a much wider body with proportions that provided a bigger female birth canal that enabled the birth of children with the bigger skulls. His height of 1.65cm was not that different from that of Homo erectus of about 1.55m, but it was the width of his body that was the important factor and not his height.

This bigger and wider body allowed further increase in the skull, which made the Neanderthal even cleverer and more versatile at tool making. But it appears that the 1400cc size skull of the Neanderthal was not enough to accommodate the bigger brain required for further abilities. The pressure was always there for a bigger brain. So the skull started increasing in size to 1500cc and then to 1600cc. But then the skull once again became too large for the female birth canal. Serious birth problems started arising.

At this stage two important parallel paths of evolution seem to have taken place.

The first change was that the body of the Neanderthal could not handle the new dimensions of the skull and this led very quickly to problems of the female at birth and to his eventual extinction.

The second path of evolution was a dramatic one that signified probably one of the biggest positive changes concerning the creation of the human race. This was the reduction in size of the neurons in the brain.

Towards the end of the Neanderthal's era, due to the pressure to reduce the size of his skull, the neurons of the brain started to become smaller. The smaller neurons had a further advantage as the speed of transfer of electrical signals from one neuron to another increased. The speed of response also increased due to the fact that the neurons were closer together.

-.-.-.-.-.-.-.-.-.-.-

The reduction in size of the neurons enabled the size of the skull to be reduced down to between about 1300cc to 1500cc. This slight reduction in the skull size improved the birth rate survival by a large factor.

The smaller but more efficient brain led to the evolution of Cro-Magnon. We can thus see an explanation as to the disappearance of the Neanderthal as well as to the reduction of skull size, which led to the virtually simultaneous appearance of Cro-Magnon.

There is now a tendency for some scientists to accept that Neanderthal was a completely different species that disappeared without any linkage to the Cro-Magnon. I consider this issue a bit further down in my book. Needless to say that such changes of views due to some temporal opinion or finding abound in the History of Science. I mention such examples in Chapter 4. Unfortunately some people treat such transitional changes in scientific thinking with passion and creed.

At the moment scientific opinion is split over this issue. So there is good justification for me to adopt the traditional thinking that Neanderthal was the predecessor of Cro-Magnon. Anyway this is only one example concerning my Hypothesis. If eventually it is proved that Neanderthal was indeed not our predecessor then the Neanderthal example that I present cannot be used to support my Hypothesis. It will not mean that all the other examples presented in this book are not valid. Just one example fewer.

This applies to all the examples I present in this book. If any of them are proved to be wrong this does not falsify my Hypothesis. If only one of the examples that I present in this book is proved right then my Hypothesis stands.

This explanation presented below concerning the disappearance of Neanderthal and the appearance of Cro-Magnon is completely different to that normally preached by most Darwinians. Their belief is that Cro-Magnon who was much cleverer and more able at tool and weapons making, outfought and eventually ousted the Neanderthals. My proposal is that fighting did not come into it. Neanderthal disappeared due to Anatomical and Physiological considerations. Cro-Magnon survived and dominated because of the sudden reduction in the size of the neurons of the brain with enormous positive implications on the further evolution of the human race.

It has been suggested that the smaller skull presented a problem as far as Darwinian evolution is concerned. This is because according to Darwin you can only have improvement in a species for survival. The smaller brain presented an apparent retrogressive step in evolution, which is against Darwin's Theory. However the explanation that I present here is in accord with Darwinism, as the smaller skull has led to a more efficient and more powerful brain. It is the power of the brain that has to be considered as evolving rather than the size of the skull.

Unfortunately many experts have always regarded the size of the skull as the criterion for ability to think and make things. My proposals completely oppose such a premise and furthermore my proposals can be regarded as agreeing fully with Darwinian thinking.

Cro-Magnon presented a big improvement compared to Neanderthal, so much so that it had been suggested that Cro-Magnon actually killed off Neanderthal as a competitor in his fight for survival of the fittest. Neanderthal was simply no match to the new "genius". However with my proposal this killing did not have to take place. Neanderthal simply disappeared because of his large skull and the associated birth problems. After all, the fossils of that era do not show significant numbers of skeletons damaged due to the owner of the skeleton having been killed.

In fact some recent findings in Portugal indicate that Neanderthals and Cro-Magnon co-habited. They found skeletons of both in the same cave and they even found skeletons with shared features of Neanderthal and Cro-Magnon.

The emergence of one species from another should normally leave the previous species unaffected, unless there is strong competition for survival, e.g. in the sharing of food and other resources. This can be confirmed for example in the peaceful coexistence of 20,000 species of grasshoppers.

In the case of the transition from Neanderthal to Cro-Magnon the first one disappeared almost simultaneously as the appearance of the second. Thus it was a transition from one to another and not just simply the emergence of a new one that could have co-habited with the old one. But one has to remember the fact that Neanderthal was already a disappearing species. From the 30,000-40,000 that they numbered at their peak, they only numbered about 2,000 at the time when Cro-Magnon emerged. Thus it is much more plausible that they interbred, resulting in the dominance of Cro-Magnon and the disappearance of Neanderthal.

It is also a much more refined explanation than to visualise that Cro-Magnon went round to all the remote parts that Neanderthal lived and simply killed every single member of the species in record time. Don't forget we are talking about 2,000 creatures living in an area more than double the size of Europe. Even finding them would have been an achievement. Anyway there was no reason for Cro-Magnon to exterminate them because they posed no threat to him and 2,000 creatures over such a vast area could hardly have posed serious competition for food or other resources.

-.-.-.-.-.-.-.-.-.-

The more powerful brain of the Cro-Magnon did not just guide him to design better tools and weapons but also gave him a new dimension to life. He started to have imagination and creativity. Wall paintings and engravings have been discovered in many caves, largely in Spain and France, which were made some 30,000 years ago. In some of these caves, mineral pigments mixed with animal fats were used to draw multi coloured animal figures.

Cro-Magnon paved the true road to Modern man.

The size of the neurons of the Cro-Magnon continued to decrease with three very important positive evolutionary trends. As discussed above, the first of these trends was the higher chance of survival at birth due to the smaller skull, and the second was the improvement in his brainpower due to higher speed of response between the smaller neurons and also due to the fact that the neurons were closer together. The third reason was the higher number of neurons that could now be compacted into the skull. As the number of neurons increased the brain became more and more powerful.

However, it appears that as the brain got more powerful, there were further significant advancements to it. And all this seems to have taken place in a very short period of time. This is something that Darwinian evolution has never been able to explain. Even the most ardent of the Darwinians have a certain degree of scepticism when it comes to the final development of the brain. They know that the chances of the human brain developing to its present state in such a short interval of time through a number of random mutations are not very high. For some people this is simply bizarre, to put it mildly.

It is difficult to visualize how randomness could change a virtual beast to an artist with imagination and creativity, within a period of 5,000 or 10,000 years, which corresponds to about 500 or 1,000 generations using our previous analogy.

-.-.-.-.-.-.-.-.-.-.-

Some recent studies in DNA structure have indicated that there appears to be no link between Neanderthal and Cro-Magnon. Such studies lead to the deduction that Cro-Magnon simply appeared 35,000 years ago out of nowhere. For many people that's an absurd conclusion. Unless Cro-Magnon was somehow planted on earth (an assumption rejected elsewhere in this book), then he must have evolved from another species and Neanderthal appears to be the only candidate. There is no continuity of fossil records to link Cro-Magnon to any other line of descent.

I am not going to uncompromisingly dismiss the results of such investigations but one has to remember that we are going through a transitional period concerning the finer details of the DNA and the general significance of the DNA code. I personally believe that the present model of the code is too simplistic and cannot at the moment answer all the questions concerning genetics. One question that comes to mind is the relationship of the DNA structure of identical twins versus that of other siblings. It is one thing to be able to provide an identity of a person through the DNA, and another to be able to reconstruct a person fully from such information. I appreciate that the code provides billions of combinations but is that enough to construct such a complex structure as the human body including not just appearance (including small details such as moles) but also behaviour such as reflex and instinctive reactions and knowledge that a baby must have in order to survive such as the suckling of milk from its mother's nipples? Furthermore as I say in other parts of this book I believe that there may be more information concerning our behaviour that is genetically acquired. There may even be an aspect of personality carried within the genetic code. I have tabulated some personal observations that indicate the correctness of such a possibility. I think that the science of the DNA structure is still in its infancy.

The so called junk DNA reminds me of the story of the watch repairer who returned to the owner of the watch not just the watch but also a small envelop containing a few nuts and bolts with the comment "these were put into the watch by mistake". Is it really junk DNA or DNA that needs further investigation?

It is acknowledged that the chimpanzee has about 99% of its DNA identical to the human DNA. But the DNA structure of the mouse is not that very different. And neither is that of the pig or the rabbit. I am mentioning this to emphasise the fact that there is still a lot more that we need to know about the DNA structure in general.

Some scientists have recently cast some doubt on the findings of the DNA studies in Germany suggesting that the accuracy of the equipment itself is suspect but also that the quality of the DNA sample has to be looked at with a great deal of suspicion.

I still maintain however that we need to know much more about DNA coding before we can adopt such findings and reject the traditional view that Cro-Magnon was the descendant of Neanderthal.

-.-.-.-.-.-.-.-.-.-.-

Returning to the previous discussion in human evolution, the next development was the increase in the number of synapses or connections between the brain cells. This dramatic increase in the number of synapses associated with the further relative increase of the size of the cerebrum is probably the stage at which Modern man appeared as the next stage of development after Cro-Magnon. **It is probably the biggest single improvement from beast to man.**

The relative size of the various parts of the brain is extremely important. For example the size of the cerebrum (the part of the brain responsible for the functions of thinking, evaluating and decision making as well as containing the intellectual aspects for planning and creativity) is virtually zero for most animals. For the apes, the size of the cerebrum is a very small part of their total brain size. The bulk of the brain of the apes forms part of the cerebellum, which is associated with the motor actions or instinctive and sensory processes that are necessary for their survival. The cerebellum mainly helps to maintain balance and posture.

The size of the brain of animals normally varies according to the size of the animal. Thus as the size of the horse has increased during the last few million years, its head has also increased accordingly to be able to hold the requirements of the bigger body. But all that has grown is the horse's cerebellum not its cerebrum.

For modern man the cerebrum occupies 80% of his brain. We should not forget that the cerebrum is responsible for intelligence and reasoning. Of the one billion neurons that the apes possess, only around 40 million form part of the cerebrum, whilst for modern man with a total of about one hundred billion neurons, over eighty billion neurons form part of the cerebrum. Thus from the point of view of number of neurons, the relative brainpower of man compared to apes is at least 2000 times greater. This means that the apes' relative brainpower is only a tenth of half of a per cent (0.05%) of that of the average human.

Furthermore, man's neurons are much more efficient in their function as they are smaller and working at higher speed and with some cells having as many as 10,000 connections or synapses to other cells. The total number of potential connections between all the neurons in the human brain is much bigger than the total number of particles in the whole universe.

Just to emphasize the enormity of the numbers of neurons in the brain, an average of 20,000 neurons must be formed per minute (yes, per minute) during the nine months of pregnancy. Can you imagine this being created by pure chance?

To start comparing the relative power of the brain function between man and apes, to include speed of transfer of electrical pulses as well as the number of synapses and their potential number of connections is such a monumental task that I would rather leave it to those who are in a better position than me to evaluate. I am confident however that such an evaluation would indicate that the ability of the human brain is many thousands of times or even millions higher than that of the apes.

The next stage of development for man was the acquisition of feedback, and the ability to talk. The process of feedback enables man to compare and evaluate things before making a decision. He therefore acquired judgment. These two functions must have happened at about the same time around 12,000 years ago. The reason for reaching this conclusion is the sudden appearance of the early Sumerian civilization, the first permanent settlement of Modern man about 12,000 years ago. This is the time that man started the domestication of animals and the cultivation of crops such as wheat and barley.

Speech is not a matter of simply developing the right shape of tongue and vocal cords. It is much more a matter of the brain. To speak, one needs to develop a language. The information concerning speech is controlled by the temporal lobe (Broca's area) in the brain. Obviously Darwinians will tell us that all anatomical requirements for speech were developed over a period of time by a large number of random mutations. Further random mutations would have been required to link the speech organs in the mouth with the temporal lobe.

The majority view is that even the Neanderthals used some form of speech some 100,000 years ago. But obviously it depends very much as to what we mean by speech. Some scientific work has indicated that even the wild putty-nosed monkeys use structured sentences with syntax to communicate between them. On the other hand, a tribe discovered not so long ago in the Amazon rain Forest, are not using proper language. The Piraha tribe cannot count and they have no sense of time. They find it easier to hum and whistle to communicate between them. So, even the term speech is not so easy to define. The time that speech was actually established as the major means of communication is therefore even more difficult to assess.

As to the time that is required for the random mutations to occur in order to acquire the relevant organs in the mouth and then link them to the relevant part of the brain in order to speak and also recognise speech, the Darwinians simply skip to the next page as if this does not concern them.

But this is a fundamental issue. **Mathematical randomness does not allow for such positive developments. This is against some very fundamental scientific principles as discussed elsewhere in this book. One cannot throw the Second Law of Thermodynamics out of the window, just to say that mathematical**

randomness has created order out of disorder! This is absurd! I find it amazing that other Physicists don't get up and shout about this madness!

As Einstein said: "God does no play dice".

The only possible method through which a complete and integrated speech system could have been developed (and not simply develop a number of individual parts that are eventually linked together), is through the process that I am describing further on in this chapter.

-.-.-.-.-.-.-.-.-.-

Having considered some plausible stages of the evolution of man from the Australopithecines let us look at some further anatomical issues.

The large snouts possessed by modern man's predecessors have been replaced by much smaller ones, since modern man does not have to depend on smell as much as his predecessors. Similarly with the large powerful jaws as well as the shape and size of the canine teeth.

The various stages of evolution from the early Australopithecines to Modern Man involve hundreds of anatomical improvements. It is estimated that we share about 99% of our DNA with our nearest living relative the African chimpanzee. But the 1% different genes correspond to a few hundred genes. Starting from the point of bipedal motion (standing on two feet), and ending with the most phenomenal mechanism, the human brain, the positive changes involved could not have possibly taken place in such a short interval of time just by simple chance mutations.

Some anthropologists have estimated that there are more than 300 distinctively different features between man and his nearest relatives in the African jungles. One should not forget that man is the only creature that laughs, smiles, cries in commotion, etc. Man also uses imagination and planning; he possesses creativity and ambition. He loves with passion, or with unparalleled parental feeling; he hates for inexplicable reasons.

Man's superior ability is not just limited to his brainpower. Man's dexterity is infinitely superior to that of any other creature. One only has to see the fine works of art (through a powerful microscope), created on the cross section of polished human hair to appreciate the extent of man's ability to control the movements of his hands and fingers.

The most dramatic changes have taken place during the last few thousand years.

If we were to compare the evolution of the horse from the eohippus of 60 million years ago we will see that there have been very few changes apart from size. The eohippus a small browsing animal, the size of a dog, with four toes has developed through various stages over 60 million years to the large grazing animal with a large middle toe and a single leg bone. Obviously his head has become bigger to be able to respond to the additional requirements of controlling all the motor actions of a bigger body. In 60 million years it has not reached the stage of speech, tool making, planning for the future, reading a book, designing a computer or flying to the moon.

This of course applies to all other animals (and plants) that have evolved during the last 600 million years.

The most obvious examples are those of the African chimpanzee and the gorilla our nearest relatives. Why is it that the chimpanzee and the gorilla have not changed in any way during the last 10 million years, apart from some recent observations that the chimpanzee uses tools to a limited extent? We do not know when chimpanzees acquired such skills and abilities; probably millions of years ago. But the time is not relevant. But would one classify this as a really major evolutionary step compared to the achievements of man?

Could the reason for this stagnant situation be the fact that the chimpanzee and the gorilla did not achieve bipedal motion? Is it because without the need to walk on two feet, they have reached a stage of development where there is no further pressure for improvement in order to survive? But then what about the penguin?

How can one explain that the chimpanzee and the gorilla have evaded the randomness of mutations for millions of years?

The duckbill platypus of Australia has remained virtually unchanged for over 300 million years. The platypus has a peculiar appearance and for most people it gives the impression that it cannot possibly last for too long, as it seems unable to fend for itself or to fight off any predators. If one were to be asked to rank creatures in order of possibility of survival, the platypus would definitely not rank in the top few thousand creatures around the world. The platypus however did not need to change over these millions of years and it has not changed.

The dragon fly, one of the most flimsy creatures that we can see around us, has remained unchanged for around 300 millions years.

How can one explain that in one line of descend there have been thousands of mutations whilst in a virtually identical line of descend the number of mutations is almost zero?

It is important to appreciate the significance of the word randomness. **Randomness cannot be selective. It is utterly unscientific to even ponder over this issue.**

-.-.-.-.-.-.-.-.-.-

In the above discussion I used the term "pressure" for needed changes towards the achievement of required features or abilities. This in many instances could be equated to "exertion of pressure" to achieve or what has been achieved. The reason for this will become more apparent when we consider a possible model as to how these changes can become genetic through modification of the DNA code.

This is what I believe is the primary force behind the marvellous accomplishments listed above that occurred during the last few thousand years. Of course the question remains as to how this has happened. How can the pressure of "needing" something to be improved, to "it is now done"?

Let us first accept temporarily that there is such a mechanism that can achieve the desired objective. If we accept this premise, then surely the above explanation as to how things have happened in such a short period of time becomes clearer or easier to accept.

Let me stress once more that there is no other alternative explanation, apart from accepting that another superior Power (divine, interstellar or other) has come to earth and guided these changes. Darwinian evolution stands no chance in even starting to explain how such a large number of positive steps have taken place through sheer randomness. This is simply absurd.

So we return to the premise that a force that can cause these changes in such a short period of time does exist.

Before I start discussing the origin of such a changing force, I have to reiterate that I am not going to be dogmatic on this issue. It is only a proposal. Unlike my previous premise where I say that a force must exist, and I am not worried if I am called assertive on this issue of the existence of the changing force, I have to accept that other people who are more knowledgeable in the field than myself may come with a more refined solution as to the origins of this force.

To be able to examine this possibility we have to consider the function of a human cell. It is well known that all the genetic information required is stored within the DNA of the cell. This cannot be disputed by anybody. It is an established fact. Any modifications to the DNA structure can happen through one or more mutations. For these changes to be passed on to the next generation these changes must be made within the sex cells.

Within the cell, apart from the DNA (Deoxyribonucleic acid) there is another nucleic acid the RNA (Ribonucleic acid). RNA is very similar in structure to DNA. One of the main functions of the RNA within the cell is to take part in the synthesis of the proteins that a cell produces. The protein synthesis in the cytoplasm (the part of the cell outside the nucleus) is controlled by molecular messengers, which are sent to the cytoplasm from the nucleus. These messengers are called messenger RNA (m-RNA), and the messages they carry are transcribed according to instructions from the DNA.

In order to synthesize a particular protein, information from the DNA has to be transferred to the m-RNA through a process called transcription. After the transcription, the actual synthesis of the protein follows, from the information carried by the m-RNA.

The synthesis of a protein in the cell is very carefully monitored and particular proteins are only made if they are needed. If a certain protein is not needed then the production of the specific m-RNA is aborted during the transcription process. The information concerning the presence or absence of a specific protein in the environment is transmitted to the DNA by other regulatory proteins.

There are other regulatory mechanisms within the cell. For example if a certain substance, say a sugar, is not present in the environment, then the cell will not generate the enzymes needed to utilize this particular sugar. Similarly when a certain substance is abundant in the cell, then the cell will stop making it. In such a case it will also stop making the necessary enzymes for its synthesis.

Thus we see that there are plenty of control mechanisms within the cell to ensure the proper function of the whole body.

We can therefore assume that there might be a possible control mechanism through which information can be gathered within the cell on the requirements of the human body according to changing environmental conditions. I will return to the discussion of such mechanism further down. In the mean time I will like to consider some examples concerning the inheritance of genetic characteristics.

-.-.-.-.-.-.-.-.-.-

In the prologue of his book "The Living Planet", the most well known Naturalist in the United Kingdom today, David Attenborough says:

"Langur monkeys from the warm plains (planes) were able to move up into the chilly rhododendron forests and gather leaves and fruit by doing no more than to develop slightly thicker coats to keep themselves warm".

So according to David Attenborough, the langur monkeys were under pressure to have slightly thicker coats and they got them! But that is exactly what my hypothesis is all about!

And that is exactly what Darwinian evolution is NOT about.

Darwinian evolution would work as follows. "There were some monkeys that through some random mutation acquired a coat that was slightly thicker than others. And those monkeys happen to be close to the colder regions. And those monkeys managed to survive in the colder regions. And those monkeys passed on their genes to their offspring". If it sounds too "Biblical", that's because it is!!!

But how can you have random mutations simply because the monkeys needed to have thicker coats? Randomness cannot be by design. Why didn't some chimpanzees in the African jungle acquire these thicker coats instead? Or some koalas in Australia, or some opossums in America?

I am sure that David Attenborough supports the Darwinian explanation of events. The explanation he has given, which by sheer coincidence corresponds to my hypothesis is a slip of the pen. However one wonders, does he support the Darwinian explanation of these events, subconsciously as well as consciously, or does he have any doubts about it just like so many other people, or is it a Freudian slip?

Bearing in mind that such Freudian slips exist in abundance in his books, one can start wondering as to the significance of these slips.

For example on the same page, referring to people that recently started to live high up on the Himalayan heights, he says:

"As people moved up the valleys, they too began to respond to the new conditions. Unlike other animals, they did not have to depend solely on bodily changes to protect themselves from the cold. With the level of intelligence and the skills that are the particular possession of humanity, they are able to make warm clothing for themselves and build fires. But they could not construct a device to help them deal with the dearth of oxygen in the air. That could only be dealt with by physical changes in their bodies. And change they did. Today, their blood contains 30 per cent more corpuscles than that of people living at sea level and is in consequence able to carry more oxygen per litre. Their chests and their lungs are also particularly large, so they are able to take in more air with a single breath than a lowlander can."

Is this a Freudian slip again or is there something more to it than meets the eye?

I could not have described these changes to the Himalayan people better myself as support to my Hypothesis. David Attenborough described it perfectly: The mountain people needed something and Nature gave it to them. Completely contradictory to Darwinian Evolution. Absolutely 100% in accord with my Hypothesis.

In fact I confess that the first time I came across this case, I thought that this was one of the best possible proofs of my Hypothesis. People could not have bothered to go and live in these high altitudes more than a few thousand years ago, if the conditions were absolutely intolerable. But obviously this happened progressively over a period of years. The higher they went and the longer they stayed there, their bodies changed accordingly to respond to the environment.

I estimate that the migration to these inhospitable heights started about 5,000 years ago as a maximum. So they have acquired these characteristics in just 5,000 years or about 250 generations.

There were a number of mutations necessary to achieve these characteristics. One mutation to increase the number of corpuscles in the blood, one mutation to increase the size of their chests, one mutation to increase the size of their lungs, and so on. And all this in just 250 generations.

For Darwinians to continue to insist that these mutations occurred randomly, after all the evidence to the contrary might be considered as an affront and insult to human intelligence.

David Attenborough is not the only writer that falls into a Freudian slip when it comes to Evolution. Alan Mann in the Encarta Encyclopaedia under "Human Evolution" states:

"It is likely that the increase in human brain size took place as part of a complex interrelationship that included the elaboration of tool use and tool making, as well as other learned skills, which permitted our ancestors to be increasingly able to live in a variety of environments."

Thank you Alan, for confirming my Hypothesis.

-.-.-.-.-.-.-.-.-.-

Let us consider the case of a white person who exposes himself to strong sunshine for a few hours on the beach. The body's reaction to this exposure is to try and produce melanin on the skin (unless the person is an albino or suffering from some other condition that his body does not create melanin). Melanin increases the rate of absorption of the useful part of the sun's radiation, which in turn provides additional protection to the body. The result of this increase in melanin is the tanning of the skin. Equally important is the fact that the tanned body reflects the harmful Ultra Violet radiation.

The body cells undertake the function of creating the melanin for the requirements of the skin. This melanin production will continue as long as the body thinks that there is a requirement for further melanin to be produced. Thus after a few days on the beach the white person will become more and more tanned.

Let us now assume that this white person started to live in a country of permanent sunshine and that he spends quite a bit of his time sunbathing by the beach. Within a few weeks or maybe months the white person will become more and more tanned. In some cases where the person is genetically predisposed (such as Greeks, Italians or Spanish) he can become as brown or black as some black people. I am sure that some of you have already seen people like that, on the Mediterranean beaches.

The cells will continue to produce the melanin as required by the body. If this sunbathing continued for many years the person will actually change his skin colour in a semi permanent way. Thus his body will continue to be tanned or black throughout the winter months in the absence of the strong summer sun. This is because his body cells will continue to produce enough melanin even in the weaker winter sun.

There is nothing controversial in what I described above. There are many people in Mediterranean countries that fall into this category. Assume now however, that this person had a child before his long term of sunbathing and a child after the long term sunbathing. Would there be any significant difference in the amount of melanin on the skin of his two children? Probably not! But there is a high possibility that the colour of the skin of the two children will be significantly different. According to modern Darwinians the answer is definitely not.

But how does one then explain the darker complexion of the Mediterranean people (such as Greeks, Italians or Spanish) compared to the much whiter Northern Europeans? Surely not another random mutation!

And how is it that as one ventures further south towards the Equator people become progressively darker and darker. For example why are Egyptians generally darker than Greeks but not as dark as the Sudanese? Why are the

Moroccans darker than the Spanish but not as dark as the Nigerians? Why is it that the Norwegians and the Danes are (on average) much whiter than the Greeks, the Italians or the Spanish? And what about the Indians? The fact that this is genetic can be confirmed at birth. There is no way this can be a case of survival of the fittest as one cannot visualize a situation where at any stage in the development of the Europeans and the Africans there has been an instance of survival because of the colour of their skin!

My premise is that the cells can be trained to produce melanin for long periods of time and that this trait can somehow be passed to the sex cells and eventually become a genetic characteristic. Thus I am making the bold proposal that the skin of our white friend's successors will become darker and darker until one day it will become black. I am not saying that they will acquire all the features of Negroes (curly hair, etc) but their skin will one day after many generations become black. As mentioned above the Australian Aborigines have dark skins but their hair is not curly.

The situation described here is not very different to that of the experiment carried out in the early seventies where scientists trained some planarian flatworms to respond to light when placed in darkness, as described at the end of Chapter 2. Through these experiments it was shown that this acquired trait was passed onto the next generation of flatworms.

A few years ago, a small minority living in North West Ethiopia, who called themselves Falashas, claimed that they were Jewish. They had been practicing the Jewish faith for many centuries. They claimed they were the descendants of Jewish settlers in that part of the world, a long time ago. Whatever else one can say about Israelis, there is nothing bad that one could say about the way they look after their own. Bizarre as this claim may have been, the Israelis thought that it merited further investigation. Israel is well known for its advanced scientific approach in various subjects including genetics. They carried out a full investigation into the DNA structure of the Falashas and found that their claim was indeed justified. This was a surprising outcome, as the Falashas had all the facial and body features of the African Ethiopians, and had no obvious facial trace of the Jews. The suggestion that they acquired these features through marriage (or interbreeding) with Ethiopians was rejected because Jews do not allow marriage outside the Jewish faith. The Falashas in particular are very strict about this; they simply do not allow marriage outside their religious community. But most importantly, the DNA structure did not show signs of such blood mixing with Ethiopians.

Historians place the settlement of these Jews in Ethiopia at around 200 B.C. Thus it seems that people who should have looked Jewish have managed to allow the Ethiopian environment to make them look like most Ethiopians in about 2,200 years or about 150 generations. Are the Israeli Genetic Scientists wrong? I frankly doubt it. I rather feel that my proposals are correct and that the Israeli Scientists simply have provided evidence to confirm my proposals.

The case of the Falashas was the subject of strong political discussion and controversy at the time both within and without the Knesset. The Israeli Government and the Israeli people definitely did not approve of 3,000 black Ethiopians coming to live in Israel. The policy of Israel on refugees is clear. They are all welcomed (indeed encouraged to come and live in Israel) as long as they are Jews. That is why there was a lot of pressure to ensure that the blood of each individual was tested and proved to be a genuine Jew before entering Israel from Ethiopia. The tests were carried out under the highest level of scrutiny and under the supervision of some of the best Scientists in Israel.

So what we have in this example is a confirmation of the suggestion I made above that in some occasions, after a number of generations, people can acquire features as dictated by the environment.

I can provide two more examples from my immediate family. We are Greek Cypriots and most of my family look typical Mediterranean – not as dark as the Egyptians but darker than the average Norwegian. There are however two exceptions. One of my mother's sisters is much darker than all her other brothers and sisters. There were nine brothers and sisters in total that reached sexual maturity and reproduced offspring. Three of the four children of the darker aunt of mine are much darker than any of their cousins. Furthermore, nearly all the grandchildren of my darker aunt are darker than any of their second cousins.

The second example in my family concerns one of my sisters who got married to another Greek Cypriot who lived for fifteen years in Nigeria before getting married to my sister. During the many years in Nigeria the colour of his skin darkened. He has become much darker than the average Greek Cypriot, the colour of his skin being closer to an average Indian rather than a Greek. They had three children whose complexion is much darker than any of their first cousins. They subsequently moved to Cyprus where they have been living for over twenty years. Their complexion is still darker than any of their cousins.

-.-.-.-.-.-.-.-.-.-

My premise therefore is that genetic characteristics can be inherited not just through random mutations but also through another process that is related to the way that the relevant RNA provides the information on the type of protein to produce. Obviously this information has to form part of the DNA. The way that the information is passed on to the DNA is what I am proposing in my new hypothesis.

It is at the moment believed that there is only "one way" communication between DNA and RNA, viz. the DNA allows the copying of a section of it onto an m-RNA, but that there is no way that RNA can make changes to the DNA. But as mentioned elsewhere in this book, experiments have shown that DNA modification can be effected by some viruses. As also discussed above, in another experiment, planarian flatworms passed onto their offsprings the ability to detect light.

This is basically all I am suggesting: that DNA can somehow be modified through other means apart from randomness. If this was to be accepted and then further investigated then all the problems associated with the Darwinian Theory of evolution will immediately disappear as the answers will be provided by the Theory of Duality of evolution: Evolution through random mutations and Evolution through RNA-DNA interaction.

This has led me to postulate the existence of an RNA whose function would be similar to that of a shift register in Electronics. This Counter RNA would simply count the frequency of production of a substance, such as melanin, and then decide accordingly as to whether there has been enough frequency of production to give a signal to modify the DNA.

I can visualize the reaction of some people who would say that such a mechanism has not been observed and such a Counter RNA has not been detected. All I can say is that, "I agree! But have we specifically searched for either of these?"

As a corollary to my argument I would like to mention the converse argument that "changes produced in an individual's phenotype by the environment cannot be passed onto the offspring". Or that "Darwin wrongly thought that habit and the effects of use or disuse could be inherited". Where is the scientific proof for these arguments regarding all the examples that I mention in my book or Darwin mentions in his? All you will hear is negative proof viz. "It is well known", "It is generally accepted", "Nobody disputes this", etc. Well I dispute it! Just give me the proof, not the statements!!!

So if there is no scientific proof for the above statements then there is justification in further examination of my proposal.

To any reservations concerning the ability to count, I can think of various processes that could be used as a counting device. Some of these could use a binary system such as used by all digital computers.

I can also mention that there is at least one plant that can count. This is the Venus flytrap whose leaf closes when an insect touches its sensitive hairs, twice within a period of twenty seconds. If the insect touches the hair only once, the leaf does not close. So the Venus flytrap can count at least to two. But not only it counts it also has a timing device. This particular example is examined further in Chapter 10.

-.-.-.-.-.-.-.-.-.-

I would like to illustrate the functionality of the Counter RNA through an example.

All tissues require a supply of sufficient amount of oxygen in order to survive. In the case where the minimum amount of oxygen required is not available (hypoxia), it can cause irreparable damage to the tissue. However the cell responds immediately and amicably as soon as hypoxia appears. It triggers off a series of emergency measures. The HIF (hypoxia-inducible factor) protein takes over control of the several compensatory activities of the hypoxic response.

The activity and stability of the HIF are regulated by the prolyl hydroxylases enzymes.

HIF is normally inactive. It is only called upon to act in the case of hypoxia. In such a case HIF gets close to the nucleus of the cell where it turns on certain genes. These genes immediately start producing the relevant proteins that compensate for the deprivation of oxygen.

HIF is found practically in all cells of all organisms from the most primitive to humans.

What I am suggesting is that the Counter RNA counts the number of times that the HIF is called upon to act in situations of this nature. If it is found that this is repeated regularly then a signal is given to modify the DNA. I am not going to speculate on the details of the mechanism of modifying the DNA nor on the minimum number of times required for the signal to be given to the DNA. These will have to be properly researched.

The word count could obviously be associated with frequency, i.e. the number of counts per second, etc. during a particular interval of time.

What I am putting forward is that once the individual's DNA has been modified then the DNA instructs the HIF to get close to the nucleus of the cell either on a regular or a permanent basis where it turns on the relevant genes. These genes then produce the relevant proteins that compensate for the deprivation of oxygen.

It is possible that the permanent change of the DNA may have further signals initiated in addition to the HIF getting close to the cell nucleus. Thus for example it may give the relevant instructions to create larger lungs and chests as well as producing higher concentration of blood cells.

My proposals fit the observations perfectly. Thus most people that climb up on a high mountain manage to survive the effects of reduced levels of oxygen (hypoxia) without too many problems. However, when they try to do some physical exercises they cannot have the same stamina as the Sherpa that have lived on the Himalayas for many generations. This has actually been proven in practice through the work of many scientists. The longer one lives at high altitude the better he gets acclimatised and the better his responses at these altitudes. This is precisely what climbers who propose to climb Everest do. But once again they will never be able to have the same stamina as the Sherpa.

The Sherpa have had their DNA modified through the Counter RNA. This has been achieved over many generations. The Counter RNA keeps a register of all such activities from generation to generation and recording all counts on the DNA. Thus the Sherpa DNA deals with hypoxia as a natural consequence rather than as a series of emergency measures. That is how the Sherpa have managed to acquire relatively larger lungs, higher concentration of blood cells, etc. This is my Hypothesis.

HIF is only one of several proteins that are classified as transcription factors. It is estimated that there are about 2,000 transcription factors. But they have not all been investigated.

It should be noted that the final solution adopted is not necessarily the one used by the transcription factors. Thus in the case of hypoxia the transcription factor calls upon the services of the HIF as a temporary measure. The Sherpa however have adopted a different solution, which is the bigger lungs, the higher concentration of blood cells, etc. The way that the final solution is adopted is something that has to be investigated.

As support of my proposal that the DNA can be modified it is worth mentioning that the DNA has already been modified to incorporate around 2,000 transcription factors. I consider this as another example supporting my Hypothesis. In fact this is a very important example indeed.

One thing is certain: this procedure cannot be based on randomness, as argued throughout the pages of my book.

What the Counter RNA provides is the creation of a record on the DNA that there is a requirement for some change to combat a particular external or environmental pressure. The value of this record on the DNA increases from generation to generation as the Counter RNA continues to increase its count, until a solution is found that permanently modifies the DNA. Thus the DNA of the Sherpa has been modified to produce bigger lungs, etc.

Transcription factors must exist to cater for many other such eventualities or shortcomings. I mentioned earlier the case of the toughening of the skin of the foot sole of people that do not wear shoes, especially those that live in the jungle. I suggest that if research was carried out, it would confirm that a transcription factor equivalent to HIF exists that toughens up the sole.

The fact that there is a requirement and that a transcription factor exists, it does not necessarily mean that a permanent solution has been found or can be found. For example the toughening of the soles of the foot has not become a permanent feature. People of African origin born in London and subsequently wear shoes, do not have tough soles.

On the other hand the arctic bear had a necessity to keep warm in the freezing conditions of the arctic. To achieve this, it has developed a layer of nearly 15 cm of fat as well as a fur that is one of the most effective heat insulators known to man. This is now a permanent feature of the arctic bear – not of other bears. Can you see how the Counter RNA has helped to achieve this?

The suggestion of the Counter RNA even though a conceptually nice idea it is only a Hypothesis and until some proof is found for its existence it will remain as such. Other Scientists who work in this area more than me could look into alternative mechanisms. It would not surprise me if such a Counter was in fact situated within the DNA. This would obviously simplify things as it eliminates the need of a mechanism to transfer this information from the RNA to the DNA. After all, the DNA is controlling most of the relevant processes and is continuously updated on the outcome of the processes through a feedback mechanism.

In spite of the enormous amount of research in this area of science there are still many unknowns. For example I mentioned above the regulatory processes within a cell. But the work that has been done on these processes has been mainly on microorganisms. Very little is known about the regulatory processes in cells of higher organisms, even though, they are believed to be similar in many respects.

It should also be stressed that the structure of ribosomal RNA, one of the most important RNA's is not yet fully determined. Similarly there are a large number of protein molecules whose function is unclear. Furthermore even though the termination of an RNA chain is well defined, there is still a problem in understanding a chain initiation.

I mention these gaps in knowledge concerning the structure and functions within the cell, simply to emphasize that one cannot be dogmatic and reject my hypothesis on the grounds that there is no evidence of such a function within the cell.

I have said above that it is believed that the transfer of information from the DNA to the RNA is strictly unidirectional. However I have also said that it has been recently established that certain viruses that can cause cancer in animals can transfer information from RNA to DNA using an enzyme with the descriptive name "reverse transcriptase".

So if these viruses have managed to transfer information from the RNA to the DNA why should we have such a big qualm or apprehensiveness about our own cells developing a way to achieve such transfer?

-.-.-.-.-.-.-.-.-.-

In a very important series of experiments, during the eighties, two immunologists the Australian E Steele and the Canadian R M Gorczynski demonstrated that it is possible for mice to acquire immunity from some diseases, that were acquired by either of their parents during their own life time.

These series of experiments have indicated that far from all arguments, Lamarckism lives in the immune system. Unfortunately the "neo-Darwinian clique" once more got together to try and discredit the work of these renowned immunologists, even though the results of their experiments have been repeated, on a number of occasions.

These experiments in themselves prove the validity of my hypothesis, without any further supporting evidence. But of course the corollary is not necessarily true. If the results of these experiments are shown to be unacceptable, they do not throw my hypothesis out as non-valid. My argument still stands until another hypothesis is proposed that explains all the unexplained phenomena in a more comprehensive and scientifically acceptable way.

-.-.-.-.-.-.-.-.-.-

During 2006 some Australian scientists observed that new born lambs do not mind eating grass with a very high salt content. These lambs were the offspring of parents which were fed on such grass, for a few generations. Normally sheep would not eat grass with high salt content unless they had no other alternative. But these lambs did not discriminate such grass from any other pasture. It was obvious that their genes had been modified within a few generations to be able to do that. Some sheep were banned on the beach on a small island on the Orkneys in Scotland 200 years ago. They had nothing to eat but sea weed. They had access to fresh drinking water. After 200 years of evolution these sheep can only eat sea weed. They would not eat fresh grass except as a small substitute of their main diet. This is fast evolution at its best.

-.-.-.-.-.-.-.-.-.-

Thus the premise that "there is no way that RNA can make changes to the DNA" does not hold in the above cases. Well if it does not hold in the cases above, why should it hold for other cases, including the modification of the sex cells to ensure that any changes are passed to the next generations?

I am quite confident that in the near future further evidence will be found to support this notion. The extensive evidence I am providing in this book is sufficient for me to make this bold prediction with confidence.

I am wondering how the strict neo-Darwinians explain such events.

-.-.-.-.-.-.-.-.-.-

The main aim of this book as specified before is to give a number of examples where Darwinian evolution could not possibly by itself provide information as to how a characteristic of a creature has evolved.

I hope that through my suggestion that an alternative process may be acting to provide a type of evolution that is controlled directly by the organism or creature itself, will help elucidate some of these unexplained situations.

Otherwise, how could one explain that it took Australopithecus Africanus one million years to produce some crude tools, whilst in another million years Homo Erectus managed to evolve to the stage of hunting in groups, and yet in 35,000 years modern man has managed to go from flint scrapers and rough wooden batons to computers, heart surgery, space travel, atomic energy, to name but a few of the marvels the human brain has developed in such a short period of time.

In fact I would go as far as suggesting that even the evolution of Australopithecus Africanus to Homo Habilis cannot be fully explained through random mutations as it would have required such a great number of them. Similar remarks can be made about the evolution of all the creatures that preceded Neanderthal.

The gradual evolution of Neanderthal however can be the first true instance where the idea of random mutations fails completely and utterly. Especially in trying to explain the change of size of his skull, as discussed earlier on in this chapter.

Maybe one should consider the first missing link specifically concerning the human race. The oldest fossils found which probably correspond to our true earliest ancestor were those of the Ramapithecus. The latest dating techniques put his age at about 12 to 14 million years ago. The next fossils found that appear to be a further development of the Ramapithecus are those of the Australopithecus Africanus who lived from about 3 million years up to one million years ago. So there is a big gap of about 10 million years where we have no fossils at all in this line of evolution.

How is it that the Australopithecus has appeared out of nowhere and in a relatively short period of time has acquired its particular characteristics? Possibly the Counter RNA could come to the rescue and provide the answer to that missing link. One could make the assumption that the reason we could not find any fossils of that period was because there are so few, as the number of the predecessors of Australopithecus were so few. But suddenly the Counter RNA takes over and the resulting species is much more successful and it thrives in rather a short period of time.

One could go even further and suggest that the early stages of life on earth viz. 600 million years ago when organised life suddenly appeared, may provide the first example of DNA self evolution. Or perhaps it was the first instance when one could apply the Theory of Duality of evolution: Evolution through random mutations and Evolution through RNA-DNA interaction.

After all, the rate at which life developed on earth 600 million years ago was quite phenomenal. Basically from just some single cell algae (which were on earth for more than a thousand million years without any change whatsoever) we suddenly have highly structured cells such as sponges, shrimps, clams, starfish, worms, etc. And this happened within just a few million years.

Immediately after that, life seems to have mushroomed on earth. Life literally exploded at that stage. Another mystery that cannot be easily explained through sheer random mutations!

-.-.-.-.-.-.-.-.-.-

Let us look at another example concerning the human race that appeared to be highly problematic to Darwin and freely admitted that he could not provide an answer through his Theory.

The human race is the only species amongst land mammals that is almost hairless. When and why did humans lose their hair? Let us see how my hypothesis provides an explanation to this unique characteristic of the human species.

My suggestion is that the loss of hair is a direct consequence of the desire of men and women, to achieve higher pleasure during sexual intercourse. Higher pleasure is achieved when there is closer contact between the bodies of man and woman. Having sex with the two bodies covered in hair is like having sex with a woolen blanket in between the bodies of the two people having sex.

The pressure imposed upon them to achieve this higher degree of pleasure, had a direct impact upon the protein manufacturing processes in their cells. After all, there is very little in life that creates higher exertion whilst simultaneously giving such high levels of pleasure to men and women than sexual intercourse.

Basically the cell was accepting instructions to ensure further and further pleasure. This was recorded by the Counter RNA and when eventually the count reached a certain value, it created the conditions to make suitable changes to the DNA.

The fact that men and women still have some hair on parts of their bodies could be due to functional reasons. For example the hair around the genitals for women could have played a big role from the hygienic point of view years ago. The hair under the armpits can absorb sweat. This can also be extended to apply to the hair around the genitals for both men and women. The hair on the head (which incidentally does not normally interfere with getting higher pleasure during sexual intercourse) keeps the head cool in the summer through the evaporation of sweat and also keeps the head warm in winter.

This functionality of the hair can also be interpreted as a further support of my hypothesis. In other words the hair on these parts of the body was not lost, because it had a specific function and use. Furthermore there was no reason for it to disappear as it did not interfere in the increase of pleasure during sexual intercourse.

This is an incredible achievement. It would have required a very ingenious designer to come up with such a virtually perfect solution.

-.-.-.-.-.-.-.-.-.-.-

One could cite another similar example at this stage. This is the fact that man alone prefers to carry out sexual intercourse with the males and females facing each other. This position is the most comfortable one for humans because of the position of the female vaginal angle. This has changed significantly during the course of evolution. The female vaginal angle has swung forward by a phenomenal amount compared to the other primates.

During sexual intercourse, man enjoys enormously touching and fondling the breasts of his female partner as well as simultaneously kissing her. Women obviously enjoy that too. It is however not so convenient to achieve this increased pleasure from a posterior entry. So the Counter RNA got to work. As the vaginal angle moved further forward (through the work of the Counter RNA), the frequency of face-to-face intercourse increased with consequential increase in sexual pleasure. This must have taken many generations to achieve. But at least it did not require randomness to be involved. There was a requirement and the achievement was through physiological and not mathematical means.

I should stress here that there have been instances of reverse evolution (similar to the occasions where people were born with tails) where a woman's anatomy is such as to be more comfortable for her to have sexual intercourse from a posterior entry. I have personal experience of this. I had a girlfriend with such a strange anatomy. Apart from that she was an absolutely gorgeous normal blond English girl.

The Grand Master Designer achieved the ultimate in human pleasure attainment.

-.-.-.-.-.-.-.-.-.-

Man is the only primate without a bone in his penis. Many people's reaction would be "what a pity!" "How an earth did we lose this wonderful facility where one would have a permanent stiff penis?"

How can my hypothesis explain such an apparently retrogressive step in evolution? As far as I am aware Darwin did not mention this particular topic.

Fortunately (or unfortunately for those of the older generations) the answer is simple. In spite of the required stiffness of the penis in order to achieve penetration, the main function of the penis (apart from reproduction) is not to penetrate but to help achieve sexual pleasure. With an insensitive bone in the penis the pleasure attainable is limited.

As in the case of the human hair, the message received by the cells was "give me more pleasure" and the Counter RNA responded admirably and amicably. Suitable adjustments were made in the DNA to ensure that more sensitive glands that maximized sexual pleasure, steadily replaced the insensitive bone.

One other aspect of the modification of the penis through the Counter RNA is the fact that the human penis is by far the biggest of all primates, far bigger than that of the chimpanzee or the gorilla. Why has this developed to such magnitude? The first reaction would be to think that the bigger penis satisfies females more than males, so it would not appear that our hypothesis could enlarge the penis.

However it is an inherent characteristic that man gets more pleasure when he feels that the woman is getting more pleasure during intercourse. So in his effort to give more pleasure to the woman, that gives him even more pleasure in return, man has developed a bigger and bigger penis. Furthermore the bigger the size of the penis, the higher the number of sensitive glands that produce pleasure. So there are two forces at work aiming for the same ultimate result.

For the same reason women have developed the experiencing of orgasm - the only female creatures to do so.

Do you think that Darwinian evolution could have possibly created these characteristics for man in such a short period of time (in just a few thousand years) just through sheer random mutations?

One should not forget that even 30,000 years ago, Cro-Magnon was fully covered with hair just like the apes.

One further comment might possibly be of interest to a number of readers. If Counter RNA modified man's penis to be the biggest and the most sensitive why has it not made it the most enduring in activity? Why are men deprived of its use so early in their lives? Why do they have to turn to Viagra? Why didn't Counter RNA take care of that fundamental feature?

The answer to that is rather straightforward. The Counter RNA is only of use if the results of its modifications are passed on to the offspring. After the time that the penis reaches its period of inactivity, not too many children can be created.

This obviously provides another supporting factor for my hypothesis. For if this was not the case, i.e. if somehow man managed to overcome the ageing problem and have an erection when he is say 80 years old, through evolution, then this would have been contrary to my hypothesis.

However, this may be a good point to be born in mind by the geneticists. It might be much better to modify the relevant genes to ensure that an erection would be achievable throughout man's active life. It will be infinitely preferable to stimulants.

-.-.-.-.-.-.-.-.-.-

The following example has probably the most far-reaching consequences of my proposal. It is concerned with over endowment of the human species. I am referring especially to the human brain.

According to Darwin, a species can evolve to the stage where it can survive. It cannot develop to a stage over and above its basic requirements for survival. Wallace the co-inventor of Natural Selection agreed on this point. His point of disagreement with Darwin was the over gifting or over endowment (a term first used by Wallace) of the human brain. Wallace suggested that the human brain must have developed through another separate process, which as he said was unknown at the time.

As is well known Wallace objected strongly to the application of the theory of evolution as invented jointly by Darwin and himself, to the human race. In fact there was very strong correspondence exchanged between the two of them concerning this point. Furthermore Darwin himself freely admitted that there were a number of aspects of the human race that could not possibly be explained through his theory. One such example that he mentioned more than once is the case of the absence of hair on the human body as discussed above.

I feel fairly convinced that if Wallace was alive today and he were to study my hypothesis, he would have regarded it as a possible solution to the problems that cannot be otherwise explained.

The human brain has reached a colossal magnitude, far more powerful than is needed for our survival. All that a human basically needs to survive is a brain that enables him to make simple weapons to keep his predators away and kill enough game to feed him. All additional abilities of the brain were created as a bonus through the use of our ubiquitous Counter RNA.

The "pressure" to have more food on a more frequent basis has ensured that humans have developed enough brain capacity to learn to live together in groups and start growing wheat and barley – one of the biggest developments in the History of the human race. All this could not have been achieved through random mutations. It needed much more than simple chance. It needed a facility to design to perfection. It needed the facility to plan ahead.

Then the human brain developed further to the stage of being able to design and build, to read and write, and so on. This development happened a few thousand years ago. There is no evidence to show that the man of today is cleverer than the Roman road builders of 2,000 years ago, or the Greek Philosophers of 2,500 years ago, or the Egyptian Pyramid builders of 4,000 years ago.

But why the over endowment of the human brain? How was it created and what was the call for such a creation? If the man of today is not cleverer than the Pyramid Builders of 4,000 years ago, it means that this over endowment of the brain was actually created before the time of the Pyramid builders.

My answer is rather tentative but still with some scientific justification. As man started to create and plan ahead, the brain realized the limitations of its own functionality. It needed to improve much more than its ability at that stage, otherwise it would not have been able to deal with the continuously expanding range of requirements and problems that it was called to resolve. So the brain instructed itself to improve over and above its own current needs in order to be able to tackle more difficult problems that it was going to face in the future, problems of planning ahead.

It is only recently, in the twentieth century that the human brain has been really taxed beyond its limits; and this possibly only by some really clever scientists. The fact that only a handful of people can understand some of the scientific theories proposed by some of our cleverest scientists illustrates my point. It is only people like Einstein and Hawking that could possibly have taxed their brains to their full capacity. I don't think that the Pyramid builders or the Greek Philosophers of 2,500 years ago or even Galileo or Newton had phased such a situation.

Could such a tremendous development of the brain be produced through random mutations?

Of course one could say that if we had billions of years and probably thousands of billions of people to ensure enough random mutations it might have been possible to produce some improvement, but to reach the present level of "virtual perfection", well the only suitable phrase that comes to mind is "no way".

If my hypothesis was to be proved wrong or without foundation, there is no way that the Darwinian Theory could provide the answer. All I can think of as a possible alternative is the arrival to earth of DNA already developed from other planets that had billions of years of evolution and populations of trillions of people, via comets or meteorites. If we accept the notion of an expanding universe, and we accept the fact that there are millions of other planets ahead of us not only just in time but also in space, as our solar system and galactic system are expanding outwards, we possibly do come across DNA from the more advanced planets.

The notion however of spacemen visiting earth and seeding earth with their advanced DNA is a notion that I find difficult to accept, as it contradicts so many fundamental laws of Physics.

-.-.-.-.-.-.-.-.-.-.-

As I mention elsewhere in this book many scientists subconsciously accept the fact that environmental pressure can produce changes to an organism. One can observe this in their writings or their speeches. We saw earlier on in this chapter in the books of the most well known Naturalist in the United Kingdom today, David Attenborough, how he subconsciously presents cases of evolution under pressure within a very short period of time.

It might however appear surprising to observe that this happens so frequently even in Darwin's writings. For example in his Origin of Species, referring to the finches in the Galapagos Islands, Darwin quotes:

"All the finches have rather dull plumage and are very similar except in the size and shape of the beaks. Those species that have adapted to seed-eating have large, heavy beaks, those that eat insects have small, sharp beaks, and so on."

Notice that he says that those finches that have adapted to seed eating have large, heavy beaks. This phrase basically infers that after the finches adapted to seed eating, they developed large, heavy beaks. But this is completely contrary to Darwinian Theory. Darwinian Theory clearly specifies that first the finches acquire the large, heavy beaks (through random mutations) and then and only then do these finches with the large, heavy beaks, do they start eating seeds.

My hypothesis clearly specifies exactly what Darwin subconsciously said above, viz. the seeds are there, the finch wants to eat them, the finch therefore develops the right type of beak to be able to eat them much more efficiently than with an ordinary beak.

Throughout the book, I am discussing further instances taken from Darwin's writings where this subconscious element creeps into his work. For example in chapter 8, I scrutinize again his references on finches that changed their characteristics to take advantage of available resources on the Galapagos Islands.

-.-.-.-.-.-.-.-.-.-

Evolution of Technology

One of the first tools used by man was probably flint stones that were used to skin their animal prey. Flint stones are a hard rock with a glassy surface. They are usually found in sedimentary rocks such as lime stone or chalk. It splits easily into thin, sharp splinters when struck by another hard stone. Depending on the shape of these splinters they are called blades or flakes. The oldest specimens of flint stones that were used as tools are approximately 750,000 years old.

Flint stones break up so easily that they also exist as sharp stones in a natural form. The breaking of larger ones to form smaller thinner ones can occur due to sudden changes in temperature. Their shape and size are absolutely random. Man used these naturally occurring splinters for many years before realising that they can improve upon their shape to make them even sharper. They achieved this by simply hitting them with another stone or even with another flint stone. It does not take a genius to learn quickly how to make really sharp blades out of these splinters.

These were the first technological designs of man – the beginning of Technology.

Thousands of years later they added these sharp flint stones to a piece of wood thus forming an axe or a spear. The conversion of tools to weapons drastically revolutionised their lives. They could now defend themselves against predators and they could also hunt their prey with their newly discovered weapons.

This technological development could only happen because of the much more powerful brain of man compared to its predecessors. The pressure to survive ensured a much bigger and more powerful brain to help him design those primitive tools and weapons.

It took hundreds of thousands of years before further discoveries were made by those primitive people with their limited mental capacity. This is how we can best describe them now. But compared to all the other creatures around them, they could be simply described as geniuses. Amongst the discoveries they made was the creation of sparks using flint stones from which they could light a fire. Someone probably noticed that this happened when two flint stones are struck together whilst trying to make sharper or thinner flint stones (flint stones were used as a means of lighting cigarettes, etc. even up to the mid 20th Century in Europe).

During the last few hundred years thousands of inventions drastically changed our life styles. It is worth noting that the impact of inventions has a cumulative effect. Frequently one invention coupled with another can lead to a much furthered improved item compared to what each one of the two inventions can produce on their own. This additive phenomenon has resulted in the sophisticated machinery of the 20th century. Thus the steam engine was mounted on a truck and this truck put on rails to create the train running on railway lines. The wheel which is probably one of the biggest inventions of all time played a primary role to achieve this development.

If we compare however a car of the 1920's with that of the models of the early 21st century the improvement is simply staggering. Thousands of small inventions have changed the early simplistic design into the modern technological achievement with computerised controls, satellite navigation systems, etc. But these thousands of improvements were created by millions of scientists and engineers working in different countries on various projects not necessarily involved with improving a particular car. Most inventions can offer advantages to a multitude of applications.

There may be groups of Engineers and Scientists who work together on a specific project such as for example improving a particular model of a car. These people continuously look around into the inventions of other people and see what invention they can incorporate into their own project. The improvement of a model of a car every year is the result of the cumulative work of millions of Scientists and Engineers around the world.

Nothing is left to chance. Every single component is tested thousands of times over prolonged periods to ensure that it is suitable for the proposed function for which it is destined to be used.

Using this methodology, every year we have further and further advancements in all aspects of technology. It is worth mentioning that the computer used in the 2006 models of some cars is more than 10 times as powerful as that of the Apollo 11 that took man to the moon in 1969.

Would an engineer even dream of creating or adopt a randomly designed item? Strange as it may seem, I was such a person. I tried to use randomness to create something functional and useful.

During the late 70's I was in charge of a project where the aim was to introduce Computer Aided Design in the design and manufacture of packaging including plastic bottles. Whilst working on this project I came up with the idea of using randomness to create new shapes of bottles. This basically consisted of allowing the computer to produce the coordinates of a few points between the top and the bottom of the bottle. These points were then joined together smoothly to produce a new shape of a bottle that was randomly generated. Perhaps as a matter of completeness I should say that some of these designs were then downloaded into a numerically controlled milling machine that produced a mould for blowing a bottle. Thus we had in our hands a sample of a bottle within minutes of it being designed on the computer screen. This was the first system in the world where a bottle was designed within a computer and then a computer controlled the making of the mould to reproduce the bottle.

However I soon realised that the new designs were anything but practical or aesthetically pleasing. They were just shapes. In fact I could describe most of these designs as monstrosities. We therefore had to work out constraints on the positions of these randomly generated coordinates and also introduced various other design guidelines that helped us produce acceptable designs.

The results of this exercise were actually a starting point for me to make me think that if I could not get practical and aesthetically pleasing shapes of bottles on the screen through randomness, how is it possible that randomness managed to create such practical and advanced features in all the living species alive today?

This is what let me to start thinking about an alternative to evolution through randomness.

As I say elsewhere in this book "virtually all organs are perfectly designed on all species".

We should also bear in mind that the complexity of live organs whether plant or animal is infinitely greater than that of man-made tools or machines. As an example, the most advanced modern cars consist of just a few thousand components. The human brain consists of tens of billions of neurons.

Furthermore whilst we understand everything about the functions of all components of the car, we only have a rudimentary knowledge of the structure and functionality of the human brain. The more we learn the more we realise how much more there is to learn.

I mentioned above that it is impossible to create acceptable shapes of bottles that have been randomly generated. One might wonder if we allowed our computer to generate many millions of designs whether we would find some acceptable ones amongst them. I have to confess that this must be theoretically possible. But one has to remember that even the word randomness is in dispute when it comes to generating random numbers through the computer. This is because there must be a starting point for a computer program to start generating random numbers. Thus to start our system to generate random numbers we have to seed it – we have to give it a number as a starting point. This number is obviously not random unless we create another process which creates pure random numbers which in itself is a fallacy. Because of this, experts do not refer to it as a random number generator but pseudo number generator.

We had the random shape generator running on a few terminals linked to the main computer and we had operators looking for a pleasing shape for a long time. We never produced one, unless we introduced various constraints.

At one time I had a grand idea that one could generate music through a random generator that might have been aesthetically pleasing. However following my experiences of creating random shapes I had to put an end to that idea without even trying it.

Be that as it may, I doubt it very much if anybody even contemplated to allow computers to create random designs of any equipment or machinery.

Throughout the History of mankind there has only been one tool that was randomly generated. That was the shape of flint stones. Even these were soon improved to make them slimmer and sharper and thus more effective.

Obviously I am excluding from this statement tools that were used to defend themselves, such as branches of trees as these were used by other animals such as chimpanzees. I also exclude the fact that naturally occurring phenomena such as waterfalls can be used to create electricity. Waterfalls are randomly created but their utilisation is harnessed through careful application of advanced technologies.

The obvious question that arises is "why is Nature so generous towards its own creations viz. all the living and yet so tight-fisted towards all the technological manifestations?"

I propose that the answer is that our basic philosophy is wrong. Nature does not allow its living species to improve through randomness but through carefully selected criteria and guidelines.

-.-.-.-.-.-.-.-.-.-

I have mentioned a number of times in this book that Darwin's assertion was based on the assumption that the number of offspring brought to life is much greater than those that reach sexual maturity for reproduction. In the case of the second process that I am proposing in my Hypothesis where the Counter RNA takes over control in the improvement of the species, numbers are not of great relevance.

Let us for example consider the case of a species of fish living on the border of the Arctic region. As it realises that there is food in the freezing waters it ventures into the freezing water to get fed. It soon however feels the effects of the cold environment and being a sensible creature it returns to more comfortable waters. When this is repeated a number of times the Counter RNA comes into operation. The fish now can stay longer and longer in the freezing waters until it eventually acquires the perfect anti-freeze that allows it to stay there permanently. The number of fish required to achieve this is not significant. Just a small school of fish can be involved. In fact if it was a very large school, such as in the case of sardines, it would be even more difficult for the new feature to dominate in the whole school and then the whole species – unless of course the whole school were to venture into the freezing waters simultaneously. So in many ways one could say that the attainment of the new feature may be inversely proportional to the total number in a species.

Obviously this is not the only criterion of acquiring improved features. However one could start to think of reasons as to why sometimes species of very few individuals manage to evolve rapidly towards perfection. The best obvious example is that of man. Bearing in mind that when Cro-Magnon appeared, there were only about 2,000 Neanderthals alive around the World, this is clearly a very good example where small numbers are easier to follow a path of improvement through the help of the Counter RNA.

Other examples where small numbers have helped the species improve are the silver sword in an extinct volcano crater in Hawaii and the titan arum in the rain forests in central Sumatra. Both these species are some of the most beautiful flowers in nature and their numbers are extremely small. They are both enormous in size, take many years to produce a flower and then they perish soon after. Not many chances of randomness and Natural Selection.

CHAPTER 4

SUNGULARITY VERSUS DUALITY

In the previous chapter I introduced the concept of the Theory of Duality of evolution: Evolution through random mutations and Evolution through RNA-DNA interaction.

I introduced this concept to account for the anomalies observed in Darwinian evolution when examining some specific examples and I suggested that there may be more than one contributing factors towards evolution. Darwinian is obviously one of them, but that is purely random and as the illustrations that I presented show, it is difficult to account for the dramatic changes that have taken place during the transition from animal to human. Apart from the other examples already mentioned that present difficulties to the Darwinian interpretation of how evolution works, I dedicate the last six chapters of this book in presenting further examples from various types of life form that cannot be entirely explained through Darwinian evolution.

It is not unusual in Science for a new idea to complement or supplement a previous theory. There are also many examples where a new theory expands on a previous theory and gives it completeness.

Furthermore there are cases where two apparently contradicting theories that aim to explain a phenomenon have been brought together to explain all aspects of the phenomenon more fully, viz. a singular explanation being replaced by a dual. This is what I would like to expand upon below.

As what I am proposing is not the replacement of the Darwinian Evolution with my hypothesis, but that the two together provide a much more comprehensive explanation of the observed facts, this corresponds to the replacement of singularity by duality.

The History of Science shows many instances where duality has replaced singularity. One of the best illustrations is the historical modification of ideas concerning the theory of light. The Wave Theory of Light alternated with the Corpuscular Theory of Light for nearly three hundred years with inputs from, amongst others, Huygens, Descartes, Newton, Young, Maxwell, Einstein, etc.

Descartes in the early seventeenth century proposed the Wave Theory of Light. Soon after, Newton through a series of brilliant experiments provided convincing evidence as to the corpuscular (particle) nature of light. Newton's theory however was replaced again by the Wave Theory of Light following the observations of Huygens on the polarisation of light, which were later confirmed by various other experiments by Young and Fresnel. This was followed by the Theoretical interpretation of these experimental observations, by Maxwell, in one of the most intriguing mathematical treatises. The Wave Theory of Light was thus universally accepted in all its forms.

At the end of the nineteenth Century however, the old idea of corpuscles re-emerged. This followed the discovery of photo-electricity. The Wave Theory of Light simply could not even start to give an explanation to the photoelectric effect. There could not be any other explanation but that light was in the form of corpuscles (named photons) that interacted with a solid surface to cause the emission of electrons and thus generate electricity. Albert Einstein received the Nobel Prize for his work on the theoretical foundations of the photoelectric effect and the Corpuscular Theory of Light. Thus the Newtonian ideas were once again fully revived after nearly two and a half Centuries.

Nevertheless there were still many unanswered questions concerning the behaviour of light especially in regard to interference, diffraction and polarisation of light. These effects could not be explained through the corpuscular (photon) theory of light.

A couple of decades passed before a final explanation was given through the marvellous theoretical works of Dirac, Heisenberg, Schrodinger and many others who laid the foundations of Quantum Mechanics.

According to Quantum Mechanics, light can be treated as sometimes acting in the form of waves (when light does not interact with matter) and sometimes it can be treated as composed of a number of corpuscles or photons (when interacting with matter as in the case of photo-electricity).

Thus the long-standing argument that lasted nearly three hundred years had no winners and no losers either: the outcome was simply a draw. An interesting case where duality has prevailed over singularity.

-.-.-.-.-.-.-.-.-.-.-

Let us consider another example. Max Planck in 1900 postulated that energy is radiated in the form of small, discrete units, which he called quanta. The following year he expanded his theory by stating that the energy of each quantum is equal to the frequency of the radiation multiplied by a universal constant, h, which later became known as the Planck's constant. Thus the energy of a quantum is $E=h\nu$, where ν is the frequency associated with the quantum, h is the Planck constant and E is the energy of the quantum.

In 1905 Albert Einstein published his famous Special Theory of Relativity where he postulated that the energy of matter is equal to the mass of that matter times the square of the velocity of light, viz. E=mc2, where again E is the energy of a particle of mass m and c is the velocity of light.

So we have two equations specifying energy E=hv and E=mc2.

It took nearly 20 years for someone to say "Therefore hv=mc2".

An obvious conclusion? Not necessarily so! De Broglie who made this "obvious" suggestion in 1924 was heralded as a genius and was awarded the Nobel Prize for Physics. This proposal inspired Schrodinger to develop his famous equations that became the basis of Quantum Physics. Wave-particle duality is now accepted as a basic property not just of light, but also of matter and energy in general.

-.-.-.-.-.-.-.-.-

We can take another example, this time from nature. It concerns the way that migratory birds find their way as they fly thousands of kilometres. The first suggestions were that birds found their way through remembering the topology of the area over which they were flying. There were also suggestions that birds use their sense of smell to guide them on their journeys. Such proposals could not however explain how birds can find their way when they are flying over the sea or even over deserts.

Then there were suggestions that they were using the sun as a guide. But then that did not explain how they could find their way at night time. So proposals were made that they were using the sun during the day and the moon at night. But once again this did not provide a complete answer as birds could find their way even on moonless nights. The suggestion was then made that they were using the stars and not the moon. But of course this again did not provide an explanation as to how birds could find their way, when the sky was overcast.

So new ideas emerged that they were using the earth's magnetic field. They made the proposal that the birds had small magnets embedded in their heads. But then there were objections that the birds could still find their way even when there were magnetic storms from the sun, during high sunspot activity.

Eventually it was universally accepted that birds are using all these methods. They actually use any one of these as a primary guide and then they use the others as backup according to the prevailing conditions.

This example once again demonstrates that a solution that has more than one aspects associated with it can sometimes provide the true answer to the problem. Singularity failed miserably with this particular example, whilst plurality provided a much better solution.

In fact I personally would not be surprised if other methods are used by birds, of which we are not as yet aware. They might even involve some other kind of force, which science has not as yet discovered. I am saying this because there are instances where birds have found their way, where all the above methods could not possibly suffice. For example a shearwater was taken by aeroplane from Wales to Boston in the USA over 5,000 kilometres away, and when left alone it flew back to Wales where it was spotted 12 days later. This means that it was flying at a speed of around 18 kilometres per hour to achieve this time for the journey. As this is the typical speed of this bird, it must have somehow followed the shortest possible route.

These are questions that still remain unresolved and the theory of bird navigation needs to be extended, to cover unexplained phenomena such as this. Butterflies are also well known for their navigational abilities, as they travel great distances between seasons.

Birds and butterflies are not unique in their homing abilities. Many of us have heard stories of cats and dogs making their way home, after they were abandoned or lost several kilometres away from home.

-.-.-.-.-.-.-.-.-.-.-

The big question therefore arises. Is neo-Darwinian Evolution absolutely 100% correct? Or is there another further element that may modulate and sometimes even dominate the evolution of species, especially of the higher species such as man?

There can be no doubt about the validity of Darwin's interpretation in the majority of cases concerning his suggestion that random mutations have played a dominant role in the creation of the various species. This is a fact that can hardly be disputed by any scientist that appreciates the significance of mutations.

The other aspect of Darwin's theory concerning the survival of the fittest is an obvious deduction that not many people can argue against.

The weakness of Darwin's Theory, is mainly concerned with the steady evolution of a particular species from its primitive original state to a more advanced state. The vast majority of the four million plants and animals alive today have not changed for many millions of years. In some cases some species have not changed for hundreds of million of years. A typical example of this is the Australian Platypus, which has not changed for more than 300 million years. The horseshoe crab has not changed for 400 million years.

The main problem that arises with such step evolution, including the evolution of man, is a matter of time. As we saw already there simply could not have been enough time for man to evolve from the wild beast that he is supposed to have evolved. One million years is simply not enough to provide the random mutations that would have been necessary to reach the level of development of modern man, from the wild beast that lived one million years ago.

Even if we were to reduce the reproductive cycle from ten years down to five years (which may have been true in the early stages of development) this only leads to 200,000 generations from beast to human. Hardly enough time to allow randomness to provide the required positive evolutionary steps!

In this book I am trying to provide further evidence to support the view that an alternative element of evolution must have contributed to the rapid development of modern man. In fact my view is that this alternative element of evolution did not just contribute to evolution through random mutations, but absolutely dominated the evolution of man.

Even though, within this book I propose a possible origin of such an element of evolution, I am not going to be dogmatic on this point. There are thousands of other scientists in the world today, infinitely more knowledgeable than me in the area of Genetics, so I will leave it up to them to consider the problem appropriately.

The aim of publishing this book is not to provide an explanation, but to bring an awareness to the fact that somewhere there is a fallacy.

This scepticism on the limitations of Darwinian Evolution is not original. As I mentioned earlier, many scientists have tried to voice the opinion that there are weaknesses on the whole concept, especially in relation to man, but always these people have been shouted down. It is possible that other scientists have come up with similar explanations to my present hypothesis but did not dare publish their views, in case the fanatics ridiculed them.

In many instances where scientists have in the past expressed a contradictory view to neo-Darwinian evolution, they have been treated by the dogmatic members of the scientific world in a manner not dissimilar to the way that lepers were treated by some civilisations, not so long ago.

I remember a series of Television programmes on evolution on BBC2 a few years ago. The series was simply based on accepted Darwinian principles, giving evidence to support the Darwinian Theory. They actually did a very good job of it. There was hardly mention of alternative views however until the last programme of the series when a discussion took place in which a number of Scientists took part. Three or four Scientists present in this discussion, started to put an argument of scepticism concerning Darwinian evolution. The word "started" fully describes the stance of the other Scientists present and especially of the person that chaired this discussion. Every time one of these opposing Scientists started to say anything at all that appeared to oppose the Darwinian Theory, they were literally shouted down by the others present. In fact there might have even been more scientists present at that "discussion" that may have wanted to express an apparent "Anti Darwinian" thought but were too scared to do so, following the treatment of the few that had already been scoffed at.

The person that chaired the discussion simply called them "creationists" and treated their views as not worth being mentioned. This person simply separated those present into two groups: those that believed in neo-Darwinian Evolution and those that believed in the creation of man by God. He and the other Scientists supporting him did not have the vision to even consider the possibility that this other group of Scientists had anything positive to contribute. I am not saying they had anything positive to say. Maybe they didn't! The fact is though, that people who behaved like hooligans shouted them down.

I am sure that this book will receive similar treatment by many who would not even read it. Unfortunately the ratio of stereotypes amongst scientists is not very different to the rest of the people in this World. One would have expected that being trained scientists they would always be open to suggestions and new ideas to further the cause of science.

It is not unusual for scientific work of great significance to be pushed aside by many people for various reasons, frequently personal ones.

One of the best examples derived from the area of genetics is the work done by the Austrian monk Gregor Mendel who in the middle of the nineteenth century founded the principle of heredity and Genetics. Mendel did such a fantastic scientific work on the acquired characteristics of peas but his work was pushed aside as of no significance. Scientists at the time did not show any respect for his work, mainly because he was not regarded as a scientist. Mendel was a monk in a monastery. A Dutch Scientist eventually brought his work to light, many years after Mendel's death.

Scientific research is not as innocent or pure as some people are led to believe. Politics, personal interests, financial interests, jealousy and mysticism sometimes dominate scientific research.

Unfortunately throughout the History of man one can find many examples of Scientists or Philosophers who paid dearly because of their convictions. Socrates, the most famous of the Greek philosophers was sentenced to death by the most Democratic establishment of the Ancient times. He simply criticised the Athenian society of injustice and he was accused of corrupting the youth by teaching them on such issues of injustice. Galileo was sentenced by religious fanatics, to spend the rest of his life under house arrest, because he believed that the earth is moving.

Some results of scientific experiments do not get published if they appear to contravene or contradict some established scientific principle (such as the neo-Darwinian doctrine) because the experimenter is worried that he or she might be laughed at by the scientific hierarchy.

Furthermore it is not uncommon for Scientists not to publish unexplained observations or results of experiments, simply because they think that these might be based on wrong assumptions or wrong samples or some other error that they cannot account for. Such experimental results are simply put to one side for further investigation in the future. More often than not they are simply forgotten in time.

During the early part of the 20th century, the Austrian geneticist Paul Kammerer was involved in some experiments involving the midwife toad. He found out that when the midwife toad was forced to mate in water (rather than on land as these toads usually do) then the males develop nuptial pads which are pigmented. This pigmentation was apparently inherited by its offspring. Not many scientists believed that such pigmentation was genetic. When Kammerer was invited to demonstrate the results of his experiments, somebody injected Indian ink under the skin of the pads of one of his exhibits. He was so humiliated that he killed himself. Investigations that followed showed that if he had done the injecting of the ink himself he would have been more subtle about the strength of the pigmentation and he would have done it on all the specimens that he exhibited, not just on one of them.

Arthur Koestler later demonstrated that the work of Kammerer still stands and he wrote a book about the whole affair.

-.-.-.-.-.-.-.-.-.-.-

I said above that there is widespread scepticism concerning a number of limitations on the Darwinian Theory. In fact this scepticism goes back a long way to the time that Darwin proposed his theory. But most objections at that time centred on religious considerations and fewer on scientific.

However the first one that expressed doubts was Darwin himself. In his first book "The origin of species" he starts Chapter 5 as follows (my own italics) :

"I have hitherto sometimes spoken as if variations were due to chance. *This, of course, is a wholly incorrect expression*, but it serves to acknowledge plainly our ignorance of the cause of variation. In the first chapter I attempted to show that changed conditions act in two ways, directly on the whole organisation or on certain parts alone, and indirectly through the reproductive system.
It is very difficult to decide *how far changed conditions, such as of climate, food, etc., have acted. There is reason to believe that in the course of time the effects have been greater than can be clearly proved.* But the innumerable complex co-adaptations of structure, which we see throughout nature, cannot be attributed simply to such action. *In the following cases the conditions seem to have produced some slight definite effect*: shells at their southern limit, and when living in shallow water, are often more brightly coloured than those of the same species from farther north or from a greater depth; birds of the same species are more brightly coloured under a clear atmosphere than when living near the coast or on islands; residence near the sea also affects the colours of insects. Certain plants, when growing near the seashore, have their leaves in some degree fleshy, though not elsewhere fleshy.

When a variation is of use to any being, we cannot tell how much to attribute to the accumulative action of natural selection, and how much to the conditions of life. Thus animals of the same species have thicker and better fir the farther north they live; but how much of this difference may be due to the warmest-clad individuals having been favoured and preserved during many generations, and how much to the action of the severe climate? *For it would appear that climate has some direct action on the hair of our domestic quadrupeds.* **On the other hand innumerable instances are known of species keeping true although living under the most opposite climates.** *Such considerations incline me to lay less weight on the action of the surrounding conditions than on a tendency to vary, due to causes of which we are ignorant.*

… I think there can be no doubt that use in our domestic animals has strengthened and enlarged certain parts, and disuse diminished them; and that such modifications are inherited. Under free nature many animals possess structures, which can be best explained by the effects of disuse. It is probable that the nearly wingless condition of several birds, which inhabit oceanic islands tenanted by no beast of prey, has been caused by disuse, as the larger ground-feeding birds seldom take flight except to escape danger. The ostrich indeed inhabits continents but it can defend itself by kicking its enemies."

Frankly I can see more confirmations and support of my hypothesis in this first page of Darwin's Chapter 5, than I expect to get from the entire scientific world when this book is published. Just like my hypothesis he is splitting the sources of evolution into two factors: The first environmental and the second the unknown or chance factor as he calls it. And the last paragraph on the disuse of parts clearly indicates how much he believed that acquired characteristics through use or disuse are inherited.

Darwinians cannot turn round and say that Darwin was wrong on these points. If he was wrong on these points then he was wrong on the whole Darwinian theory. In such a case why are they called Darwinians and so fervently support Darwin's theory?

-.-.-.-.-.-.-.-.-.-.-

I must make some further references to Darwin's work. When referring to the human eye, in Chapter 6, he makes the following comment:

"To suppose that the eye with all its inimitable contrivances for adjusting the focus to different distances, for admitting different amounts of light, and for the correction of spherical and chromatic aberration could have been formed by natural selection seems, I freely confess, absurd in the highest degree..."

I disagree with him, because as he himself says the octopus has an eye very similar to that of the humans. So this means that the eye has had many millions of years in which to evolve to its present advanced design.

I am not suggesting that the eye followed a series of positive random mutations, and the Counter RNA did not play a part in it! All I am saying is that this is one example that nature had plenty of time in which to develop through whichever means an almost perfect tool.

As I explain in many instances in this book, my main worry to evolution is not what has evolved over millions of years, but what has evolved over thousands of years, with man, and especially the brain of man as the prime example.

-.-.-.-.-.-.-.-.-.-.-

Let us consider the chances of producing an improvement in any characteristic.

To produce a change in any characteristic, one would have to produce a change on the gene structure in the DNA chain. This is done frequently these days in various laboratories around the world, once the genetic code is identified for a particular characteristic.

Thus, Scientists could in principle, produce bigger fruit once they have identified the corresponding genes in the DNA chain that controls the size of the relevant fruit. For instance Scientists have managed to change the genetic code of some tomatoes that can be kept fresh for a longer period of time.

The DNA chain can be regarded as a very long chain whose links are of four different bases, a (Adenine), b (Guanine), A (Thymine) and B (Cytosine). Three links together form a triplet. These triplets can be regarded as a code word, where the order in which they are placed defines the code. These triplets are called amino acids. There are a total of 64 possible triplets, which can be formed by using the four letters in sequence. In fact there are only 20 major amino acids that are used in the DNA structure of the cell. The other 44 are not used.

A sequential group of these triplets forms a gene, which is the hereditary unit. The order in which the various triplets are sequenced, defines the gene. Thus one gene may define the colour of the pupil of the eye whilst another may define how straight or curly the hair would be. The number of triplets in a gene varies depending on the type of gene. In some cases this number could be over 1,000 triplets. All the cells in the body contain absolutely identical copies of the DNA. Different cells do of course perform different functions within the body.

The DNA chain is hereditary and under normal circumstances an identical copy is passed from parent to offspring.

The message within a gene can change completely, if for some reason one of the links is damaged or one of the triplets is modified to another. This random change of the code, which leads to the change of the message within the gene, is called a mutation. This is how changes can occur which lead to different characteristics of the offspring, if they occur in the sex cells. According to Darwinian Theory, mutations can happen due to various reasons but they are always random.

Scientists have tried to induce mutations under laboratory conditions by bombarding cells with different types of radiation.

The outcome has always been deleterious. For example they managed to produce flies with one eye, flies with no wings, flies with one, two or three legs, etc. They have never managed to produce flies that were in any way or form an improved species of the common fly.

The question therefore arises. Why is randomness in nature different to randomness in the laboratory? After all, a mutation caused by radioactivity, is a random mutation and cannot be any different to a random mutation caused by other random forces. The source of randomness is not relevant.

There is no scientific explanation that can be postulated to explain differences in randomness. Artificial radiation that changes a gene can only change it randomly just like natural mutations cause random changes to the gene.

To expect an improved species out of randomness is just like to expect that by opening a computer and moving a few wires randomly around, inside the computer case, we would get an improved performance from the computer, i.e.

double its memory or double its speed. Admittedly we have to imagine a computer of a few years ago, where inside the case of a computer one could see quite a few wires. These days, the inside of a computer has been streamlined so that all one can see inside the case of a computer is a small number of individual devices connected together through plugs.

Furthermore the Second Law of Thermodynamics, one of the most fundamental Laws of Physics, forbids such transitions i.e. creating Order out of Chaos. The Second Law of Thermodynamics stipulates that you cannot get Order out of Chaos (disorder) without external intervention.

One textbook definition of the 2nd Law is **"Any system that is subject to random agitations will eventually attain its most disordered condition."**

Hence if you start with a perfect creature, randomness will eventually produce the most imperfect creature. Thus randomness could change a man to amoeba – not the other way round

In fact the Second Law of Thermodynamics specifically stipulates that every action within an isolated system will result in an increase of disorder. This increase of disorder is measured through what scientists call the Entropy of the system. Just like you cannot create energy (the First Law of Thermodynamics), similarly you cannot decrease Entropy (or increase order).

Even though the DNA chain may be regarded as an isolated system, and one could argue that the Second Law of Thermodynamics could apply within its domain, one would then have to postulate another explanation for the evolution of lower forms of life from protozoa and the formation of various forms of more advanced species over many millions of years.

For, I think there is no doubt that random mutations led to the creation of many species from the original protozoa.

One could of course argue that if I accept that random mutations can produce a new species that can sometimes survive through Natural Selection why do I insist that no improvement can take place through randomness? The answer to this is that randomness can produce diversity but not enhancement to the chances of survival of a species. This can only be done thorough my Hypothesis.

-.-.-.-.-.-.-.-.-.-

Darwinians take great pains to stress the fact that the development of new species takes place through a large number of mutations over a long period of time.

By studying the DNA structure however and the way that the various types of mutation can occur one can see that a very big change can sometimes take place with one single mutation. This is possible, as a mutation can create completely new information if for example one of the links (nucleotides) in the chain defined above is removed. In this case the codes defining the gene are reshuffled and thus the gene is completely redefined and can produce entirely different characteristics. Even worse, the termination of the gene (which is itself a code made up of nucleotides) can also be modified following the restructure of the codes, in which case the information of two genes can be joined together with unpredictable consequences. This can extend to a number of genes until a new termination code is created through the new restructuring.

This could easily be the reason that a new species is formed. It is a far more drastic change than say tens or even hundreds of the more common mutations where one nucleotide is substituted by another.

The substitution type of mutation is by far the most common of all types of mutation and the one with the least consequences.

It has been recently suggested that one single mutation was responsible for the transformation of an ordinary small size lizard to one of the biggest Dinosaurs, the Tyrannosaurus Rex which was 14 metres long and weighed up to four tons.

The process of many sequential common or substitution type mutations to form a new species has fewer chances of accomplishment than a single nucleotide removal mutation.

My proposal is that species are created through random mutations (of the nucleotide removal or deletion type), and then further adjustments and improvements are made through the Counter RNA until the species achieves virtually perfect functionality. These finer adjustments could possibly be compared to a series of the common mutations of the substitution type.

One could look at the way some of the domesticated animals have evolved through careful selection procedures. Looking at some dogs these days the variability between them is so great that some people might regard them as animals of different species. These have been achieved through small very carefully controlled steps in genetic manipulation.

-.-.-.-.-.-.-.-.-.-

It would be logical for someone to pose the question "if I accept that a new improved species can be created from random mutations why do I not say that this is contrary to the 2nd Law of Thermodynamics which I stress does not allow randomness to improve a species?"

The answer is that such a mutation produces a *different* species and not a *better* species. It is up to Natural Selection to decide whether such a new species will survive or not. After a very long time we will know if the new species is an improvement on the old one or not. Obviously the Counter RNA may be called upon to improve the species after it has been formed.

CHAPTER 5

MAMMALS

The large white bear of the arctic sea has developed an elongated and streamlined shape that is more convenient for its aquatic life, compared to land-dwelling bears. It has also developed an extra layer of fat beneath the skin that provides further insulation from the cold as well as a fur that is one of the most effective heat insulators known to man. The layer of fat is nearly 15 cm thick. It has the typical flat feet of bears, but its five sharp, curved claws have been developed especially for grasping the ice and its prey. To be able to withstand the severe cold it has developed long hair between the pads as well as stiff hairs on its forelegs. Its broad front feet have been developed to act as oars when swimming. To escape the notice of predators (mainly hunters) all polar bears are white, which provides a camouflage in the white snow.

All these characteristics of the large white polar bear are necessary for its survival. It could not for example survive if it had developed its front feet in such a form that they act as oars, but had not developed the extra layer of fat beneath the skin to act as extra insulation against the freezing waters in which it swims. Both characteristics are essential for its survival. Evidently one characteristic could have developed before the other. It is obvious that its extra layer of fat had to be developed earlier as that layer of fat provided insulation against the cold weather even before it migrated to the seas.

The development of its front feet to act as oars could have followed to give it additional chances for survival. But this development could not have been very much later as its importance for survival is rather self-evident.

The question is, how could randomness provide such characteristics in a short period of time? The chances of successive random mutations to produce such required characteristics are infinitesimal. The conclusion is that the large white bear could only acquire these essential characteristics through our Counter RNA.

-.-.-.-.-.-.-.-.-.-

Male elks have ferocious fights head on with their antlers to preserve their harems. They sometimes run into each other at great speed and the sound of their clashing heads can be heard kilometres away. The size of their antlers is an indication of strength. Their antlers can get broken during such fights. In spite of these vicious encounters of their heads no brain damage results.

You can imagine two boxers in a boxing ring running from opposite ends and clashing their heads together. The results could be devastating.

But not with the elks! Their skulls have developed a cavity, which acts as a shock absorber during these clashes.

In the absence of the cavity, the elks could not fight furiously for the right to mate with the female elks. Otherwise, either or both of the male elks could die through such fights. Was this development of the cavity a random development or was it enforced through the pressure to survive? The possible course of development was that before acquiring the cavity, the fights were not as ferocious, but became more and more so as the development of the cavity was becoming more and more effective.

The chances of this happening through a random mutation are infinitesimal.

-.-.-.-.-.-.-.-.-.-.-

The most characteristic aspect of a camel's anatomy is its hump. This hump contains fat, which during the day it absorbs heat from the sun and during the very cold desert nights this heat keeps the camel warm. The hump acts like a storage heat radiator. Furthermore the hump provides a shield against the direct rays of the sun for the body underneath. This ensures a reduction in the surface area from which moisture could evaporate.

The most essential feature of a camel's usefulness to its owners in the desert is the fact that a camel can survive without water for a long period of time. The camel has evolved in such a way that it minimizes sweating, so it loses little water through its skin. The camel achieves this by increasing its body temperature by up to eight degrees. Furthermore it has developed pouches in the stomach that store water, which is released, as and when the camel needs it. A thirsty camel can drink in excess of 100 litres of water to replenish its reserves. Finally in case of emergency, the fat in the hump can be broken down to produce water that can then be used by the camel's body.

A camel's nose has a long passage, which acts like an air conditioning system. As air is breathed in, it is warmed up to ensure that any moisture in the incoming air is fully utilized by the camel's body. As air is breathed out, the air passes over a cooler section where any moisture in the air is condensed and not allowed to escape.

The camel's blood cells have been modified so that their shape has become oval (as opposed for example to the human cells which are spherical). This is extremely useful when the camel has to take in enormous amounts of water after a very long period without water. If a human were to drink a corresponding amount of water under the same conditions, the human cells would absorb so much water that they would explode. The oval shape of the camel's blood cells however allows the cells to expand by a large factor without any damage.

This is a long list indeed of required features to be able to survive the long distances that it has to walk in the hot desert with limited water supplies.

But could the camel survive a long trip in the desert if any of these features were not present? Doubtful! Once again the only way that the camel could have acquired these features is through our Counter RNA. To achieve such essential features through a series of random mutations would take a bit more than simply waiting for millions of years for them to happen. Acquiring one or two of the features would not have been enough for a very long journey in the desert. Obviously the length of the journey increased as further and further features were acquired.

-.-.-.-.-.-.-.-.-.-.-

The Star nosed mole has developed on its snout 22 fleshy tentacles or rays in the form of a star that are used to detect potential prey. These rays are probably one of the most sensitive devices for the sense of touch possessed by any creature. It can detect minute creatures a fraction of a millimetre in size. The size of the sensors on these tentacles is less than $1/500^{th}$ of the size of a pinhead. This star shaped projection is unique for moles. They have been developed as a necessary requirement to give them the ability to detect their prey in their elaborate burrows below the surface of the ground.

The burrows dug by the star nosed mole are so elaborate and enormous in length that it takes the mole up to three hours to go round checking every corner for new arrivals. As worms dig their own tunnels they find their way into the moles' burrows. When the mole has consumed as much as it can at any one time, it bites them passing onto them a substance that paralyses them and then stores them away for future consumption.

But surely the star nosed mole could not have survived in such an environment unless it could detect its prey. It cannot see its prey and it cannot smell it. So the tentacles and its ability to dig those elaborate burrows must have been developed simultaneously. Hardly the outcome of two random mutations occurring simultaneously!

-.-.-.-.-.-.-.-.-.-.-

The Antarctica seal has developed the almost perfect heat insulator in order to survive in the freezing waters of the Antarctic sea. The seal has developed a thick blanket of fat together with a wrap of fur. These seals have also developed especially large mouths that help them carve the ice to create openings to the surface when under water in order to breathe. They also developed larger eyes with higher sensitivity that helps them see in the darkness under the ice. These large mouths and large eyes are not to be found on any other seals that live outside the freezing waters.

The milk of the seal for its pubs contains 20 times as much fat as ordinary cow milk. This keeps the pubs warm plus it enables them to grow up much quicker.

So once again we have to accept that the Antarctic seal could not have developed this long list of essential characteristics whilst in the Antarctic Ocean as it could not get there without previously possessing these characteristics.

The only possible way that the seal could acquire these features is through living in the fringes of the Antarctica and venturing for small periods of time into the colder parts, thus applying pressure for development of the required characteristics. And once again it could not have acquired these characteristics through random mutations, as the unanswered question would be "why are there not any other creatures with such characteristics outside that environment?"

-.-.-.-.-.-.-.-.-.-.-

The elephant's trunk is one of the most versatile devices devised by nature to perform a large number of functions. The elephant uses its trunk to eat, to drink, to spray water or dust over its body, to fight, to pull down trees, to cut tree branches, rip off foliage, to push along baby elephants, etc. Elephants can also examine small objects using their trunks, through the use of delicate finger lobes, which are found at the end of the trunk. They also use a sucking action to pick up small objects by the trunk. By raising its trunk high up in the air, it can detect wind direction as well as any scent in the wind.

If this trunk was acquired through random mutations then it must have required quite a few mutations to achieve this versatile tool. Even if the elephant acquired its trunk with its numerous functions through a previous species, say the mammoth, these comments could also apply to this previous species.

The idea of a series of positive random mutations to achieve these characteristics is unacceptable. Thus if due to a random mutation the trunk appears, then through another mutation the delicate finger lobes appear, then the suction ability develops, then the sensitivity to wind is developed, then the sensitivity to scents, and so on. The chances of achieving these positive steps through random mutations are mathematically so remote that any mathematically minded person would discount them as impossible.

Furthermore one has to consider the number of generations through which all these positive changes have taken place. Even if we go back to the mammoth for the start of the first mutation, bearing in mind that the mammoth first appeared about 4 million years ago and that the elephant starts mating at about 15 years old (with a gestation period of nearly two years) we can see that the total number of generations available to achieve all these changes is about 240,000; not an enormous number.

What I believe happened is that the trunk itself appeared through a random mutation. Then however the pressures of survival and the pressures of making life easier have enabled further developments of the DNA to achieve the further features. The idea of the Counter RNA as a means to achieve these essential characteristics is far more appealing indeed.

-.-.-.-.-.-.-.-.-.-

The case of the giraffe is probably the one that is mentioned more often than any other when discussing the principle of evolution. This is because Lamarck mentioned it first as a typical example to describe his theory and then others followed it through to clarify why Darwin's Theory provided a better explanation of the evolution of its special characteristics.

If we assumed that Darwin is right and that one member of the species acquired through a random mutation a longer neck, then obviously this member and his offspring had a higher chance of survival. However this first mutation did not produce the length of the neck that the giraffe has today. A second random mutation was necessary to make it even longer. If this happened, this particular giraffe and its offspring would have had even a better chance of survival, on the assumption that other competing animals were not similarly endowed through a similar beneficial random mutation. This process would have had to continue for many thousands of generations to produce the cumulative effect of the longest neck in the animal kingdom.

One should not however forget that the long neck of the giraffe is not enough for its survival. Reaching the leaves at that height is one thing, picking the leaves from frequently thorny and interfering branches is another. The giraffe, together with its long neck, has simultaneously developed a very special tongue that can be up to 50 cm long. This versatile tongue possesses high elasticity whilst being very tough, and is covered with numerous papillae. It is specially designed to capture the leaves that are hidden behind thorny branches.

There have also been other arguments raised concerning the length of the giraffe's neck, for example that special mechanisms needed to evolve simultaneously to adjust the blood flow to its head to ensure that it does not faint whilst the animal is drinking water. Some biologists have demonstrated that quite a few mutations are required to achieve such a characteristic.

Since blood needs to be pumped several meters up to its brain, it has double the average blood pressure and the strongest heart in the world. The heart weighs 11 kg and is 60 cm long. Its walls can be up to 7.6 cm thick. The brain can't stand high blood pressure, so when the giraffe bends down to take a drink, its blood pressure would destroy the brain. So the giraffe has developed one-way check valves in his jugular veins, which close when the head is lowered. But there is still too much blood in the carotid artery. This extra amount of blood is taken to a piece of spongy tissue full of small blood vessels near the brain, which soaks it up.

Furthermore, the cerebrospinal fluid (around the brain) produces counter pressure to prevent capillary rupture.

So once again we are talking about cumulative beneficial random mutations. A logical independent observer who knows nothing about the giraffe or evolution would immediately say "How can you have cumulative benefits through randomness? These are two contradictory terms!" They are indeed!

For someone who knows nothing about the giraffe or evolution but knows something about the Second Law of Thermodynamics would immediately say "How can you get order out of chaos? This is impossible!"

To crown it all off, recent studies have indicated that the main purpose of the long neck of the giraffe is not for picking up leaves from the tall trees. What the studies have pointed out is that the male giraffes use their necks mainly for fighting amongst themselves during the mating season. This necking gives considerable advantage to those giraffes with the longest necks.

For whichever reason the giraffe has acquired a long neck, the explanation through our Counter RNA can help without any contradictions or randomness. All that is needed is for the giraffe to feel the pressure that a longer neck is necessary for its survival or even its dominance. The Counter RNA will take over to make the right adjustments in the DNA.

-.-.-.-.-.-.-.-.-.-

The vampire bat flies out in the evenings until it locates its victim and then lands close to it. It then hops quietly towards its victim until it gets very close to it. Victims include horses, cattle, and other big animals. It pierces the skin with its razor sharp teeth and then sucks up the blood as it flows out of the wound that it creates. The victim feels no pain because the saliva of the bat contains an anaesthetic that kills the pain on the animal. The saliva also stops the blood from clotting which is essential, bearing in mind the fact that sometimes the bat may be feeding for up to two hours from one wound.

If the victim felt any pain it would have used its tail to get the bat away. The question that arises is, "could these three essential characteristics of the vampire bat be produced simultaneously through a random mutation?"

If one of these three characteristics of the vampire bat had not developed together with the other two, then the bat could not have possibly survived in its present form as a vampire bat viz. living on the blood of other animals.

-.-.-.-.-.-.-.-.-.-

Bats have developed an echolocation system to detect insects at night. They emit an ultra sonic sound, which bounces off the insect. The reflected sound is detected by the bats, enabling them to locate the insect accurately and then they move towards the insect and catch it. To ensure a high resolution of the position of its prey, the bat is using higher frequencies than ordinary audio frequencies. The normal audible frequencies that humans can hear range between 20 cycles per second to 20,000 cycles. The bat uses frequencies between 50,000 and 200,000 cycles per second. They send out a number of bursts of sounds each second, each one sounding like a click to them. This is the reason that we cannot hear the echolocation sounds made by the bat.

Bats judge the distance from a particular object, by the time it takes for the reflected sound to get to them. As they approach the object, the time gets smaller and smaller. To improve the accuracy of locating the object, the bats increase the number of clicks they send towards it. Bats have also developed a system whereby the quality of echo defines the shape and type of body of the object that reflects the signal. Thus it can discriminate between an insect and another object nearby similar in size to the insect.

As bats normally eat their prey as soon as they catch it, they cannot send any more clicks until after they have finished eating the previous capture. To avoid this problem some bats have developed a system by which they send clicks through their noses. In addition to this development they have further developed a facility by which they can concentrate and amplify the sonic beam. They have furthermore optimised the reception of reflected sound through an elaborate system of translucent ears, with special ribs spikes and blood vessels. It sounds like bats were involved in the design of the "Amplifier and Speaker" systems much before our recent electronic era.

One particular moth has developed a defensive device against the bats echolocation system. These moths also emit an ultrasonic click, which jams the bat's echolocation signal.

The bat has however developed a further attacking technique. It uses its very long ears that are like acoustic horns. Once it locates the approximate position of the moth, the bat switches its echolocation system off. It then listens for the very low sound emitted by the slightest movement of the moth's wings. The moth can then not escape.

These "cat and mouse" complicated and comprehensive features cannot surely be the work of randomness!

There are still some questions that remain unanswered concerning the echo locating system of the bat. For example in the case of huge numbers of bats as in a confined space, how do they separate the returned sonic signal to ensure that they reject all other signals apart from their own? Some people have proposed complicated filtering mechanisms similar to those that exist in radio receivers, but again this does not fully explain the filtering out of interfering sounds from hundreds of thousands of bats that may coexist in a cave.

One species of bat has actually found a way to catch fish. Its wings start higher up on the leg compared to most other bats, so that its legs are free to catch fish. Because of this position of the wings, they can stay out of the water when the bat dives to catch a fish. Its toes have been modified to be much larger and its claws have changed to be hooked shape. Furthermore it has developed a technique whereby it can compensate for the refraction of the sonic waves as they go into and out of the shallow level of water above the fish.

Marvellous inventions through "randomness"!

-.-.-.-.-.-.-.-.-.-.-

It is thought that bats have evolved as a species to take advantage of the extensive presence of insects during the night. Birds lived on insects that they captured during the day. Millions of years ago there was no other species of animal that fed on the enormous numbers of night insects, so there was pressure for species to be developed to take advantage of the existence of night insects.

To achieve that, such a species had to be a flying species.

It is well known that there are various species of flying squirrels that in reality they only glide from higher branches of trees to lower ones. They achieve this through a membrane of fury skin that looks like a blanket. This membrane stretches between its long forelegs and hind legs. This membrane was possibly developed to enable the squirrel to improve its ability to catch additional prey and also to avoid some of its predators. I am not suggesting that bats evolved from squirrels. But what I am saying is that the flying squirrel demonstrates a possible transitional stage that the bat may have gone through to get to its final stage of flying.

Once again the pressure of developing characteristics to achieve something that a species was not originally designed for, is demonstrated clearly.

-.-.-.-.-.-.-.-.-.-

The South American giant anteater can consume more than 30,000 ants every day. It has developed very strong front claws, which it uses to dig into termites' and ants' nests, with such ease as if the nests were made of paper. It has also developed a very long snout and a long sticky tongue with which it picks up the ants and termites from their nests. The snout and the long sticky tongue provide a very efficient mechanism indeed for the purpose for which they were developed. When at work it sticks its long thin tongue in the opening that it creates in the nest, so rapidly that one wonders how these ants get stuck to the tongue so quickly and then disposed of in its mouth with equal speed. It is unlikely that the best engineers of today could have developed a more efficient form of device for this purpose. It also lost all its teeth during the process of evolution, as it did not need them any more.

Once more, one would find it difficult to understand how these characteristics could have been developed through a sequential number of random mutations. The loss of its teeth due to atrophy is also extremely important. It is much more plausible that the Counter RNA was used to develop these special features of the South American giant anteater. The Counter RNA also detected the fact that the teeth were not being used so they were gradually removed from the anatomy of the anteater. There was no reason to use additional effort to grow unwanted body parts.

There is a sequel to the comments made concerning the giant anteater. There are numerous unrelated creatures that have become anteaters in many parts of the world. In Africa and Asia for example there are several kinds of pangolin that have become anteaters. The armadillo in South America is also another well-known anteater. Even a marsupial in Australia has become an anteater. There are other anteaters mainly in Africa and South America. All these creatures have developed the above characteristics viz. the long snout and the long sticky tongue. Even birds that have taken to ant eating such as the woodpecker and the wryneck have also developed long beaks with very long sticky tongues.

The common feature of all anteaters is a long sticky tongue and they have all lost their teeth. Also they are all very slow in their movements. It appears as if though without a long sticky tongue a creature would not try to make ants its main prey and source of living. There are of course other creatures that love eating ants; even the chimpanzee as we discuss elsewhere in this chapter, loves to have a snack of ants at times. But ants are not the main food for the chimpanzee.

We should of course say that there are many creatures with long tongues, such as the chameleon, frogs, etc. which do not get involved in ant eating in a big way. As described in Chapter 7, their tongue has been developed with different special characteristics. The special tongue of the anteaters is found on creatures that live in areas where huge numbers of ants exist and ants form the main part of their diet – ants of course can be found just about everywhere in the whole world. The numbers of ants do however vary enormously between different places.

According to Darwinian evolution the anteaters acquired their long sticky tongues through random mutations. The question that begs is "why have we not seen these long sticky tongues that have been developed specifically for ant eating, on any creatures in areas where there are no huge quantities of ants"? One could say that if such creatures did exist, they could not last very long as there were not enough ants around to feed on. This may be an explanation. I believe that these creatures developed their unique tongues over a period of time. They started eating ants with whatever tongue they had previously and the pressure of improving their tongue to become a more efficient tool got the Counter RNA to work.

-.-.-.-.-.-.-.-.-

Armadillos, which live mainly in South America, have developed very strong Armour, which gives them higher protection from potential predators. This Armour covers their entire body and in some species even their tails. It is made of horn, the hard material derived from hair and bony plates, and which is formed by the ossification of the greater part of the skin.

Armadillos have an extremely sensitive nose and they can smell any prey lying quite deep under the ground. Once they detect something they put their snout to the ground and start digging at an unbelievable speed. As their snout is immersed deeply in the hole formed in the ground, they are unable to breath. In fact their system has been developed to allow them to withhold their breath for around six minutes, while digging.

Some smaller armadillos do not have their entire body covered by Armour. The three band armadillo of Paraguay has its shoulders and rump covered by a single shield whilst the middle of its body are covered by articulate transverse bands that allow it to roll up to cover its abdomen when approached by a predator.

The rolling up of shielded animals is common when their entire body is not shielded. The ordinary hedgehog is the most well known example in Europe. As soon as it feels that it is threatened, it rolls up to cover its exposed parts such as the head and the abdomen. It is very difficult to make a hedgehog unroll, until it senses that the danger is no longer present. I personally found out that if you put the rolled up hedgehog in a bath of water, it immediately unrolls and starts swimming. The pangolin of Africa and Asia, rolls up by bringing its head into its stomach and then again it is almost impossible to make it unroll until it feels that it is no longer threatened.

-.-.-.-.-.-.-.-.-.-.-

Some mammals, among them the desert oryx, which is an endangered species of antelope, vary their body temperatures, storing heat by day and releasing it at night when the desert gets really cold.

The question that arises is why such a feature was not inherited by mammals that live in colder countries, but only by mammals that live in very warm countries and especially in deserts. Thus such a feature could not have been randomly acquired.

-.-.-.-.-.-.-.-.-.-.-

Some animals in the jungle start walking as soon as they are born. Take the wildebeest for example. Within a few minutes the baby wildebeest is up and running about. It also has its eyes and ears open at birth. What would happen if it did not get up immediately and instead needed attention for a few days as many other animals do? Well it would have become immediate prey to the lions, the cheetahs, and so many other predators in its environment.

The idea of natural selection viz. that there were baby wildebeest that got up reasonably quickly and others that did not, and that eventually those that got up quicker survived, whilst those that did not get up quickly became prey to the predators, is a non starter. Simply because there are too many hungry mouths to feed amongst the predators and they would not miss a chance of a baby wildebeest lying on the ground for more that just a few minutes.

-.-.-.-.-.-.-.-.-.-

Dolphins and whales use greasy tears to wash away the salt from their eyes. Could this be the result of a random mutation? I doubt it. And I think that by now so do you. This may be a matter of comfort but it is a serious aspect of the well being of these mammals. They definitely required this feature and "prayed" for it, as the stinging effects of salty seawater could be unbearable.

-.-.-.-.-.-.-.-.-.-

Whales breathe in an enormous amount of air to keep alive whilst they are beneath the surface of the water. When it comes up to refresh the air in its lungs, it blows the old air out through its nostrils. But as the head of the whale is normally below the surface of the water, it would have needed a large amount of energy to push the air out into the water. So its nostrils have been moved on the upper part of its head, which is above the surface of the water when it is floating on the water surface. Thus when the whale expels the used air in its lungs, the air shoots upwards which requires very little energy to do so. If its head is just below the surface then we can observe the familiar fountain of water, which is pushed by the air coming out of its nostrils.

-.-.-.-.-.-.-.-.-.-

The main predator of dassies (rock hyrax that look like large rabbits) is the eagle. Before attacking, the eagle watches the dassies carefully, to ensure that they do not see it when it attacks. The eagle's normal strategy is to attack with the sun behind it so that its prey is dazzled by the strong sunshine in their eyes. However the eyes of the dassies have developed a means by which their eyes are not dazzled, they can face the sun straight on. So they can see the eagle coming and run away from the danger.

-.-.-.-.-.-.-.-.-.-

It has recently been discovered that some chimpanzees in the rainforests near the borders of Guinea and Liberia have developed into exceptional tool users. They mastered the method of cracking oil palm nuts to get their soft edible kernels. For this purpose they use one stone as an anvil and another stone as a hammer. They have even found out that a flat anvil stone is more effective. Some of the chimpanzees have become highly efficient in this technique. They even found out that it is better for them to carry the selected stones with them under a tree, rather than keeping the stones in one position and carrying the oil palm nuts to that position.

One other recent discovery was the observation that some chimpanzees chew some leaves to use them as a sponge to draw water from tree hollows. They simply put the chewed leaves in the small pool of water, allow it to soak up water for a few seconds and then pull it out and put it into their mouths where they squeeze the water out.

The most incredible discovery however concerns some chimpanzees that climb up on the top of an oil palm tree, they pull out some leaf branches that they use to pound into the soft part of the top of the tree. This softens the tough fibres, which the chimpanzee cannot eat, into a sugary mash. They then eat this sugary mash.

It was previously observed that some chimpanzees use thin sticks that they dig into a termites' nest, which the termites immediately attack. As the attacking termites get attached to the stick, the chimpanzees pull out the stick and they quickly lick away the termites on the stick.

Chimpanzees have also been observed to use sticks to angle algae from pools and lakes, which they seem to enjoy immensely.

It is estimated that the chimpanzees of this rain forest use up to about 20 different types of tools.

One wonders how different these tools are to the first tools used by the early Australopithecines. These discoveries made some scientists to start rethinking the process of development of man. The present view is that the australopithecines started using tools only after becoming bipedal and started living on the savannah away from the trees, on which they lived for millions of years before that.

It should be noted that some of the chimpanzees in the same groups that such tools are used extensively, are completely unable to use any tools at all. They simply watch the tool users at work, and some of them don't even try to use these tools even though the advantages are immediately obvious. This makes me feel that the acquisition of such tool making techniques are genetic. Those chimpanzees unable to use the tools have simply not acquired the right genes to do so.

How did the chimpanzee manage to discover the use of these tools? The answer must be that necessity is the mother of invention. One can visualize the situation where there is not much food around except of a large number of oil palm nuts. They had to react to a crisis. Their reaction was probably instinctive. The chimpanzees must have known that the inside of these nuts was edible, as some of the nuts must have cracked through natural causes, or birds may have dropped some which they cracked with their stronger beaks. They must have also been aware of other softer nuts which they cracked using their teeth. It must have been a true inspiration for the first chimpanzee that actually tried the use of stones to break the first nut. This probably happened many thousands or millions of years ago. The other inventions must have followed a rather similar pattern.

-.-.-.-.-.-.-.-.-

Most mammals instinctively suck the breast of their mother as soon as they are born. The kangaroo is however an exception. The marsupial young cannot suck the milk out of the breast of its mother. The mother injects the milk into the offspring's mouth when it clings to its mother breasts. To ensure that the milk does not enter the windpipe that could lead to choking, the young kangaroo has a relatively long larynx rising into the posterior end of the nasal passage. This allows free access of the air to the lungs whilst allowing the milk to flow freely along the sides of the larynx.

-.-.-.-.-.-.-.-.-.-

Dolphins and whales have evolved from land mammals. They obviously had to adapt to the water environment. The rear limbs have been lost altogether whilst their forelimbs have become their paddles. They lost their hair completely, as it is of little use to them in the sea. To ensure that their body is kept warm they have developed blubber, which is a thick layer of fat, underneath their skin.

Some whales including, the blue whale, which is the biggest of them all, feed on krill, which are tiny shrimp-like crustaceans. They have lost their teeth, as they are not of any use to them. Instead they have acquired the baleen, which is a sheet of horn, feathered at the edges that hung like curtains from the top of their mouths and act as filters. The whale looks for a shoal of krill, goes in the middle of it and then swallows a huge amount of water including a large amount of krill. It then shuts its jaws and expels the water out. The krill remains behind as the water is filtered out through the baleen.

To illustrate the advanced mathematical and hydrodynamic knowledge acquired (through randomness!) by the blue whale, it is enough to say that in order to create a concentrated mass of krill, when it finds a shoal of krill it goes underneath it and blows large numbers of bubbles towards the surface of the water. This produces a hydrodynamic drug that pulls the krill towards the bubbles. The whale then opens its mouth and moves upwards, whilst at the same time it swallows the water and krill.

The dolphin uses echolocation in exactly the same way as the bat, as discussed above. The dolphin using its advanced sonar system can detect prey located under the sand. This is not as easy as it sounds, as the sound waves are distorted as they travel from water through to the sand, then to the prey and then back again. The dolphin has to be able to compensate for all the secondary waves created during this process. But of course being the mathematical genius that it is (as possibly endowed by randomness!), it has no problems in precisely locating its prey.

-.-.-.-.-.-.-.-.-.-

It is believed that the domestic cat is derived from a species of African wildcat, the caffre cat. This is believed to have happened in ancient Egypt about 4,500 years ago. The Abyssinian breed definitely looks very similar to the caffre wild cat. Even though however it looks very similar, they are a different species. They could not interbreed. They have some characteristics, which are completely different between them.

The implications of this situation are enormous as far as evolution is concerned. The Abyssinian breed is supposed to be nearly the same as the breed in existence in ancients Egypt 4,500 years ago. Most experts seem to agree that the caffre wild cat simply walked into the lives of the Egyptians when it realised that it could get a free regular meal.

So how long did it take for a new species to evolve? As far as is known there was no other animal that it could interbreed. It simply evolved naturally.

But according to normally accepted evolution theory this should have taken a few million years, not just hundreds of years, if we were to assume that the Abyssinian cat that lived with the ancient Egyptians was the same as the Abyssinian of today!

-.-.-.-.-.-.-.-.-.-.-

From cats to dogs! One other example, which according to some authorities could be even more recent than the Abyssinian cat, is the Australian dingo. There are contradicting opinions about the age of the dingo. Most set it at about 3,000 years old but some believe that the dingo is a species of dog that has gone wild again since the Europeans settled Australia. This makes the dingo as only about 200 years old.

Whichever date is the correct one, one thing is for sure that the dingo is a very recent species that evolved in record time as far as Darwinian evolution is concerned. In fact Darwinians will never accept that there could be a new species created in such a short time. Darwin always spoke of very long periods viz. of millions of years or at the very least hundreds of thousands of years in isolated cases.

-.-.-.-.-.-.-.-.-.-

Let us look at another dog, the pointer. This is a breed of large dog that hunts by scent. When the pointer's nose picks up a game it will stand still immediately and its head will point in the direction of the game. This instinctive behaviour of the pointer is genetic. There are conflicting views as to the exact origins of the pointer. All the experts however agree that it has evolved during the last few centuries.

So once again we come across a non-Darwinian evolution. Darwinian evolution requires random mutations over a long period of time. A few hundred years could hardly be regarded as a long time for evolution purposes.

Mammals eating leaves and grass have a problem with digesting the cellulose contained in them. Animals chew the leaves as much as they can to break them down to as small pieces as possible, but still this is not enough. Cellulose is one of the most difficult substances to break down and digest. The digestive juices in the stomach are usually unable to deal with the digestion of the cellulose. The only way this can be done is by some special bacteria within the stomach of the animal. Even then however this process can take time.

The rabbit has invented a way to ensure that it takes in as much of the nourishment as possible. After the leaves are shredded by its incisors and ground by its molars, they are then passed to the stomach where the bacteria and its own digestive juices take over. Eventually they are passed into the gut, where they are formed into small pellets and excreted within the burrow of the rabbit. The rabbit then eats these pellets and after they are swallowed they go through the same process for the second time to ensure that no nourishment is wasted. After the second process is completed then the rabbit excretes these new pellets outside its burrow.

-.-.-.-.-.-.-.-.-.-

Some grasses have developed silica in their tissues, which have the property of wearing down the teeth of animals that live on them. This is obviously a defensive mechanism of the grasses against their predators. However these animals have developed a counteractive measure that helps them overcome this serious problem. They have developed gaps between the roots of their teeth. These gaps enable the teeth of these grass eaters to grow continuously throughout their lives. As their teeth grow at the roots, the growth is pushed upwards which balances the loss due to the wearing out by the silica in the grass.

-.-.-.-.-.-.-.-.-.-

CHAPTER 6

FISHES AND OTHER SEA CREATURES

Let us now consider some examples from various species of fishes and other sea creatures.

Lungfish are a kind of fish that can be found in some swamps and marshes in the Tropics, mainly in Africa, America and Australia. Lungfish possess gills as well as an air bladder very similar to the lung of higher vertebrates. They breathe oxygen dissolved in water, through their gills, whilst they can breathe through the air bladder by occasionally coming to the surface of the water. Thus lungfish have a dual breathing system.

The earliest fossils indicate that this group of air breathing fish were in existence nearly 400 million years ago.

When the swamps and marshes dry up during the dry season, the African lungfish get into a burrow and close the entrance to the burrow with mud except for a small opening through which air can reach them for breathing. They then cover themselves in a protective layer by secreting a mucous and aestivate - fall into hibernation. So the lungfish of the African swamps have the ability to survive away from water for a long period of time. When the rainy season returns and the marshes return to their former glory, the lungfish come out of their burrows and into their natural environment. Even if there is no rain during the following season, the lungfish can survive in their cocoons for periods in excess of four years.

The question arises: How many random mutations were necessary to acquire these highly specialised characteristics, bearing in mind the fact that in the absence of any of these characteristics, lungfish could not survive for too long in the African dry season? Without these characteristics in their totality, lungfish would be dead at the beginning of the dry season and thus there would be no fish in the swamp the following season for any further evolution to take place.

The Darwinian view that there were many types of fish in the swamps that evolved during a long period of time and due to random mutations some of them managed to develop these characteristics, that led to their survival, has to be utterly rejected. Lungfish simply didn't have the luxury of time to develop these highly specialised characteristics. These characteristics had to be developed simultaneously, otherwise they would not have been of any use to the lungfish.

Another possible explanation is the consideration of acquired instincts as this is an acceptable premise even in the Darwinian Theory – a subject treated extensively by Darwin himself. It is possible to visualise some lungfish to have entered the burrow by instinct or by design as the soil was wet and it was a more acceptable environment than the dry surface of the bottom of the lake. The idea of drilling a hole for breathing was again through a natural instinct or design for survival, as it realised that it was running short of air. This behaviour can be compared to the acquired habit of some animals such as the chimpanzee to use tools or a tree branch as a defensive weapon. Thus the lungfish did not survive by sheer chance but by a simple act of self-preservation – probably taking some action for temporary survival and hoping for the best.

Be that as it may, the lungfish acquired some characteristics in order to survive. As these characteristics had to be acquired simultaneously, or over a very few generations, assuming the conditions were not too unfavourable, the only explanation is the Counter RNA as proposed in the previous chapters.

-.-.-.-.-.-.-.-.-.-

Climbing fish are a common occurrence in a number of countries in Africa, and also in India, Malaysia etc. There are numerous species of walking fish. They usually come out of lakes that are about to dry up, and they start to move about on land. They seem to retain water in cavities connected with their gills. They have even been observed to climb up trees.

Like lungfish they seem to possess gills for breathing under water as well as an air bladder for breathing when outside water.

Similar remarks made about the lungfish also apply to the walking fish.

-.-.-.-.-.-.-.-.-.-

Have you ever wondered how some fish survive in the arctic sea at temperatures many degrees below zero?

These fish have managed to survive in these conditions because their lever creates a very effective antifreeze, similar to the antifreeze that we use in our cars during the winter months, but much more efficient in its function.

So how did these arctic fish acquire such ability? Let us assume that it was acquired through random mutations (Darwinian Evolution). There are two possibilities to consider:

    1. Such mutation happened whilst the fish were in the arctic sea

2. Such mutation happened before the fish reached the arctic sea

The first possibility has to be discounted simply because the fish would have not survived under such conditions even at the point of trying to get into freezing areas of the Arctic Ocean, without the facility of the antifreeze. As soon as it tried to get into the freezing waters it would have frozen itself to death.

So the mutation must have happened before these fish found themselves in the Arctic Ocean. Armed with this facility such fish reaching the arctic seas would be experiencing no difficulties swimming in these inhospitable waters. So any species of fish with the facility of creating antifreeze could easily drift into the freezing waters of the arctic seas.

But if this is the case there must have been species with this ability to create antifreeze through their lever in other regions of the world. Why is it that no such species has been found outside the arctic seas?

Let us now consider another possibility. Fish in the region close to the arctic seas suddenly discover available and unexploited food in the freezing waters. They can go in and out in the fringes hunting for this food. The body of the fish senses the requirement of being able to sustain longer periods within the freezing waters. The body somehow has to provide the facility to allow the fish to achieve this. The body of the fish comes up with the answer – allow the lever to produce an effective antifreeze (glycoproteins).

Certainly the body of the fish is neither omniscient nor does it possess a universal remedy for all eventualities. The fish's DNA has to give the relevant instructions to try various options until a suitable solution is found. In this respect a random process has to be assumed. However according to my proposed principle, randomness is only the beginning of this process. For the proposed mechanism assumes a learning process through which the solution is achieved through a series of processes following an asymptotic curve. In other words, every type of solution tried is used as a stepping stone towards the final solution.

Thus the Counter RNA is once again the saviour of the situation.

In fact some other species of fish in the Antarctic evolved a different method of surviving in the freezing waters. They fill their bodies with highly concentrated salt solution (sodium, potassium, chloride ions or urea) which lower the freezing point of their bodies. This illustrates that the various species of fish facing similar problems can find slightly different solutions through trial and error. All it signifies is that the Counter RNA can provide different solutions according to the specific requirements and circumstances.

In spite of the absence of evidence for such a process, this must be a more acceptable solution than the alternative of randomly generated mutations whereby the fish has to wait for millions of years (in reality it might probably require billions of years) until its body comes up with the right mutation to produce the required antifreeze.

As a matter of argument one could ponder as to why other creatures in different environments have not acquired this characteristic of producing antifreeze! Our story does not end here however. Ice fish have also adopted another useful characteristic: This is the lack of haemoglobin. Haemoglobin is the red coloured pigment in the blood which carries the oxygen around the bodies of every other vertebrate species. This is a useful adaptation as oxygen is highly soluble in the cold sea water. This also makes their blood thinner, allowing their metabolism to be slower, which conserves energy. Ice fish also have extremely efficient enzyme systems which allow them to remain active at low temperatures - their activity at 0° C is similar to that of a temperate water fish at 20°C.

There are 122 species of ice fish in the Antarctic Ocean. This large number of species within only one area of the earth surely signifies that randomness could not have had much to do with their evolution with these unique features.

-.-.-.-.-.-.-.-.-.-

The Amazon cod found it impossible to see in the dark, muddy waters of the Amazon. So it developed sonar guides like bats.

Again one could ask the question "why other similar fish have not developed sonar guides in clear waters?"

The sonar guides were obviously developed because there was pressure on the Amazon cod to be able to find its way through the muddy waters. It could survive without it but life would not have been too easy.

-.-.-.-.-.-.-.-.-.-

The hammerhead shark has developed special sensors that can detect minute electromagnetic fields that are created by the muscular movements of creatures in its environment. In fact the hammerhead shark, through these sensors, can pin point very small prey. The goby is only 15cm long and lives in burrows in the sand or mud of the seabed. As it draws water through its mouth and out through the gills, it produces an electrical signal, which is detected by the hammerhead shark. This electrical signal is so minute that would be hard for the most advanced man-made detection equipment of today's technology to detect it.

No other land animal has developed such a highly sensitive device. The shark cannot see in the dark near the mucky seabed. Without this characteristic the available feeding options of the hammerhead shark would be very much reduced. Is it not obvious that necessity determines the course of evolution?

-.-.-.-.-.-.-.-.-

It has been recently discovered that a certain species of lobster, the palinuridae, during the period when it sheds off its hard shelf and therefore its defences are weakened, it produces a sound to scare its predators. This sound is produced by a mechanism similar to a violin. The movement of its antennae results in the movement of its plectrum over a region close to the eye which appears to be like a rasp. This forward and backward movement produces a sound similar to the way that the vibration of the violin strings produce a sound when the bow is drawn across the strings. The sound made by the defensive system of the lobster is however not as rhythmic or melodious as that of the violin. After all, the purpose of this sound is not to attract but repel.

Let us consider the implications of the development of this defensive mechanism. If this mechanism was developed my pure random mutations, which is quite possible, did the lobster have the luxury of time on its side to wait for such a mutation to occur?

With the risk of sounding monotonous, I have to reiterate that my proposed hypothesis provides a much more plausible explanation as to how such a defensive mechanism has evolved.

-.-.-.-.-.-.-.-.-

During the long winter darkness in the Antarctic the generation of plankton, the main food for many sea creatures including whales ceases completely. There is simply no food for either plant or animal life during these months. Yet they all somehow manage to survive. Whether they shut down their metabolic activity or they use some internal food storage for these months is not known. The whale finds the answer by migrating to other waters far away from the Antarctic.

But one thing is for sure. All the sea creatures that depend upon plankton in the Antarctic did not have the luxury of time to wait until the "God of randomness" decided to come up with the right mutation to give them the chance for survival. Without food they would have simply vanished.

-.-.-.-.-.-.-.-.-.-

The whale shark in its seasonal travel from the Tropical islands of the Indian Ocean to the coast of East Africa, a distance of nearly 2,000 kilometres, has to travel through areas where there is very little plankton during that time of the year. The whale shark lives on plankton. Yet it manages to survive. Through some unknown process the whale shark is aware of the fact that there is going to be a deficiency of plankton ahead and optimises the expenditure of its energy. In fact it allows the seasonal currents to carry it along at very low speeds. It hardly uses any energy at all. It simply floats and lets the water currents do the work for it.

How does the whale shark know about this deficiency of plankton ahead? As it travels alone it is obvious that this information must be part of the genetic code that it inherited. Could this be something that was created randomly? Before answering the question we have to bear in mind that if the whale shark were to use even the slightest extra amount of its energy than what it is at the moment, it would simply perish.

So if we were to assume that those whales that conserved their energies survived and those that did not conserve their energies perished, then there would have been many more that perished simply because the natural tendency of the whale would have been to use its energy to get to its destination as soon as possible.

Again the elements of time and numbers become important. Bearing in mind the numbers of these creatures and the other natural reasons for not surviving was there enough time for them to evolve to their present state with the acquired characteristic of energy conservation?

-.-.-.-.-.-.-.-.-.-.-

Cuttlefish are well known for their ability to camouflage themselves in most environments. This chameleonic characteristic is very important for its survival. It is a characteristic possessed by a number of creatures.

Cuttlefish have also developed a means of propulsion that is of great significance not only for swimming away from predators but also for getting to their prey when required. Cuttlefish normally swim through the use of narrow fins that surround their bodies. They have however developed an opening behind their head, through which they can squirt water at a high rate. This siphoning results in the propulsion of the cuttlefish at a very high speed for a short distance. This additional means of propulsion provides the cuttlefish with obvious advantages for survival and one could not reject the premise that such a characteristic could have developed through Darwinian evolution (randomly generated mutation). One could visualise the situation where such a characteristic was passed from generation to generation over many millions of years.

Glass shrimp have transparent bodies, which helps them to escape the notice of their predators. These shrimps have developed bodies that are transparent to light. They also developed transparent muscles and transparent blood. They are therefore practically invisible to most fishes around it. An excellent defensive mechanism! Glass shrimps have also developed relatively thick muscled abdomens, which they contract rapidly to make a sudden escape from their predators.

In spite of these defensive mechanisms developed by the glass shrimp, it still provides the cuttlefish with a large proportion of its diet. As explained above, the rapid escape of the glass shrimp using its thick muscled abdomen is counteracted by the jet propulsion characteristic of the cuttlefish.

Similarly cuttlefish have developed the ability to see the glass shrimp, which is almost invisible to other creatures in the environment. This is achieved through a development in their eyes that enables them to see the glass shrimp through polarised light reflected by the body of the glass shrimp. This is similar to them wearing polarised glasses.

Cuttlefish can also see through their polarised light detection facility other light coloured fishes that normally reflect light from the scales of their bodies in such a way that they are nearly invisible to most creatures around them.

The cuttlefish has also developed the feature of approaching its prey by opening out its tentacles in such a way that it appears to be a floating dead weed. It addition to that it changes its colour to appear as unobtrusive as possible. In other words it has also developed the art of mimicry. It attacks its prey when it is very close to it and within reach of its striking ability. It catches its prey by suddenly ejecting its long tongue, which is nearly as long as its whole body.

Once more these "cat and mouse" phenomena are attributed, by Darwinians, to randomly generated characteristics over the millions of years that these creatures have been developing. However one has to consider the chances of such a large combination of characteristics being developed in such an environment through pure randomness.

Once more I find myself reiterating that such a theory is not beautiful. It lacks finesse it lacks elegance. It is simply clumsy and ugly.

-.-.-.-.-.-.-.-.-.-.-

Let us consider some further examples that give support to my hypothesis. I present these examples without any comments.

Fish that live in the tide-pools in the tropics can be exposed to very high temperatures. This can lead to overheating and to death. Some snails have developed special ridges along their shells that act like heat radiators. They simply increase their total surface area to act like the cooling ribs of a car radiator. This cooling effect keeps the temperature of the snails lower so that they are not cooked when exposed in the hot sun.

To overcome this problem, the dog whelk snails have evolved a white colour that reflects the sun's radiation to keep them cooler.

The giant electric eel has developed the ability to detect its prey through electrical signals emitted by most creatures due to the movements of their muscles or other parts of their bodies. When the eel approaches its prey it stuns it from a distance with an electric charge of around 600 volts. It then moves closer and eats its stunned prey. This must be a very clever animal indeed to have created electricity millions of years before Faraday, Volta and Van de Graaff.

The water monkey, a fish in the Amazon basin has developed eyes, which have split focus. The lower part focuses downwards whilst the top part focuses upwards looking for insects on the trees that overhang the water. When it sees an insect on the leaves of the tree, it jumps out of the water and catches the insect. This fish is about one metre long and it can jump up to one metre above the water surface.

The splashing tetras are also fish of the Amazon basin that have realised that their eggs have a higher chance of survival if they are placed outside the water, whilst waiting to hatch. The female jumps out of the water and lays its eggs on a leaf overhanging the water. The male immediately jumps onto the leaf as well and fertilises the eggs. When the eggs are hatched the young fish fall into the water.

Some kind of small molluscs, the Chitons, have developed eight overlapping plates on their backs, which effectively reduces the amount of moisture that can escape from their bodies. In addition to this, they can survive the onslaught of the hot sun by having developed a tolerance to the drying out process. They can actually lose up to 75% of the water in their bodies, and still survive.

Some sea weeds have increased their degree of tolerance even further. They can survive even if they lose 90% of the water in their bodies. When this happens they can actually become completely dry and crunchy. Yet when the water returns, they recover rather quickly.

The "cat and mouse" game described elsewhere in this book is quite evident amongst sea creatures. Mussels and oysters have developed hard shells, which they can close to defend themselves against a number of predators.

Some snails however have managed to overcome this defensive mechanism of the hard shells. These snails have developed their own attacking mechanism, which consists of a tool similar in effect to an electric drill. Using this tool, which is situated in their mouths they simply drill the hard shell of their prey. They then secrete some digestive enzymes through the hole. The body of the prey is turned into a thick soup, which the snail sucks out.

Stonefish are more poisonous than any other fish. They lie dormant on the seabed, perfectly camouflaged, waiting for the opportunity to strike whenever a fish passes close enough to it. It has developed a camouflage that makes it look like any other rock on the seabed.

Harlequin shrimps love to eat a certain type of starfish, which are much bigger in size than themselves and which normally lie down on the seabed. The harlequin shrimp cannot break into the hard topside of the starfish. It can however find its way into the starfish flesh if it had access to the other side of the starfish, the side touching the seabed. The harlequin shrimp has developed a technique by which it turns the much bigger starfish upside down. It simply takes the end of one of the tentacles of the starfish and lifts it up until it turns the starfish over. The reason it lifts it up from the end of one of the tentacles is to increase the leverage of lifting. The pressure for survival has enabled the harlequin shrimp to invent the principle of the lever millions of years before Archimedes.

The archerfish lives in mangrove marshes. Its main prey consists of insects that are quite abundant amongst the branches and leaves of the mangrove. It has developed a technique by which it squirts high speed water in the form of a fine jet towards insects above it on the leaves or the branches of mangroves. It has developed the technique to ensure high accuracy.

-.-.-.-.-.-.-.-.-.-

There is virtually no light at all at depths of around 1000 metres in the sea. Yet there is still life down there. The main problem that fish are experiencing down there is seeing their food. However these fish have come up with a multiplicity of characteristics that enable them to survive. These developments form very strong evidence for my hypothesis as it illustrates very clearly once again that "need is the mother of invention". As I explained elsewhere in this book, the most important element in evolution of required and desired characteristics is the need of such characteristics. This environment illustrates very clearly that the number of specialised acquired characteristics are much greater than normal environments.

The deep-sea angler has very small eyes, which are not sensitive enough to detect their food. To overcome this weakness it has developed the most sensitive listening devices. These consist of very long hairy antennae, which grow on its skin all over its body. Following from the previous paragraph, it is obvious that such antennae could only have been developed in such an environment. They are simply not needed by fish that live in the open seas at only a few metres from the surface. It is nothing to do with randomness. If randomness were the governing factor then such antennae would have also appeared on fish that live a few metres from the surface.

As prey is not very abundant in these waters, all the fish in these waters have developed powerful sharp teeth, very large mouths and large expandable stomachs. These ensure that once prey comes within the range of these fish they don't miss anything independent of size. Again the comments made about required characteristics in these environments, also apply to the development of the large mouth, powerful razor sharp teeth and expandable stomach.

In addition to these properties most of the fish in the deep seas have developed special and clever ways to lure their prey to them. Many of them use bioluminescence to draw their prey to them. The deep-sea angler has developed a pole on top of its upper lib. At the end of this pole is a device full of symbiotic bacteria, which produces the bioluminescence. Filaments hang from this device that reach as far as the entrance to the mouth. These filaments also glow. The prey of the angler sees these glowing filaments and goes closer to the mouth of the angler. The open large mouth never misses. After all the prey comes virtually into its mouth.

One other group of anglers have developed the luminescence emission device at the roof of their mouths, obviously the best possible position for it.

A further group has developed two luminescence devices, one on the top of the head and the other below its chin.

The colour of the bioluminescence of these fish is normally blue or blue green because these wavelengths are not absorbed as much as the red end of the visible spectrum. Furthermore the eyes of most deep sea-animals are more sensitive in the blue end of the spectrum. Can you see any further evidence here that the Counter RNA has been involved to produce these special characteristics?

A few species however have developed the ability to produce red light. The loose-jawed dragon fish, which produces red light, has also developed special sensitivity in his eyes to be able to see the red light that it produces. It can therefore see the prey around it, without the prey being able to see the dragon fish. This technique is similar to the Infra Red cameras that the military are using to detect movements of its enemies. This is further strong supporting evidence as to how the Counter RNA has provided its assistance to develop these specialised features.

Some species are using bioluminescence for defensive purposes by confusing their predators. Copepod crustaceans developed a system by which they create very strong light flashes. These flashes appear to serve two purposes. The first is to warn other copepods of the imminent danger, and the second is to attract attention from the predators of its attackers. This technique has been developed further where the copepod crustacean propels packs of bioluminescent materials from its body that confuses its attackers, after these packs burst with the emission of instantaneous light flashes. The story does not end here however. In a further development the copepod crustacean has managed to create a time delay in the explosion of these packs. This time delay ensures that the attackers chase the flashes while the copepod crustacean has disappeared from the vicinity.

A group of shrimps have developed a technique whereby they can produce a sticky and glowing substance. When a predator attacks them they immediately release this substance covering their predator with it. This predator then immediately becomes visible to its own predators because of the glowing substance that gets attached to it.

Do you realise that Darwinians insist that all these characteristics, that are a unique feature of the deep-sea creatures, are the result of random mutations? And worse still they do not provide an answer as to why these features have not been developed by creatures outside such an environment. Surely randomness cannot selectively choose place and time!

-.-.-.-.-.-.-.-.-.-

Let us look at other features that have been developed in the deep-sea. The reason I give great attention to this environment, is because as I said above, the difficult environment creates the pressure to create special characteristics for survival.

Most creatures have their eyes in such a way that they can see what is happening in front of them. The deep-sea squid Histioteuthis however needs to be able to see the silhouettes of its prey overhead as it crawls at the bottom of the sea. To achieve this, one of its eyes has become much larger than the other one and is situated in such a position that it can be used to look upwards. Through this larger eye it can detect moving silhouettes in the very dim light coming from the surface. The small eye is used for seeing things below.

The ubiquitous deep-sea angler that we studied above extensively has developed a rather interesting reproduction process. The female angler is more than 10 times bigger than her male suitor. Once the male has located a female (through a complex bioluminescent flashing procedure), it bites on the side of the female and attaches himself permanently into her body through his mouth. In fact more than one male can attach himself onto the huge body of the female. Once the males get attached they more or less become part of the anatomy of the female. They feed through the female's body. When the female releases her eggs, the males have to release their sperm simultaneously. To ensure perfect timing of these two processes the female takes over control and ensures that indeed the males release their sperm at the same time as the female releases her eggs. Once again such a complex feature could not have possibly been achieved through a random process.

-.-.-.-.-.-.-.-.-.-

The twilight zone is situated at around 150 metres from the surface and the creatures in this region have developed in such a way that their characteristics reflect the fact that there is very dim light present. Because of these conditions the creatures in this region have developed characteristics that reflect a "hide and seek" situation.

Within the twilight zone, where there is very little light, most invertebrates have managed to make themselves virtually invisible. The squid and the octopus for example have made themselves transparent because their bodies do not need to be as strong as those of their brothers in the shallow waters. Even in the case of organs such as the eyes, which cannot be made invisible, these creatures have developed a reflective silvery layer that breaks up the outline and their predators cannot see them.

Again within the twilight zone, some heteropods that are from the same family as ordinary snails have also managed to make themselves more or less invisible, to avoid detection by their predators. The incredible thing about these creatures is that even their external shells have become almost transparent. Also, these heteropods have acquired a rotating set of razor-sharp teeth that are carried on a long mobile proboscis similar to that of an elephant.

The eelpout, which has not become transparent has developed another technique to avoid its predators. In the presence of its predators it coils up. It brings its tail round to its mouth to form a complete circular shape. It is believed that this shape confuses its predators that take it to be a jellyfish, which they do not want to touch due to its powerful sting.

To improve its eye sight the Winteria has developed very sensitive large tubular eyes. Furthermore its eyes have been developed so that they look up towards the surface to detect the moving silhouettes of probable prey.

The Hatchetfish has developed a flat body to lessen the chances of being detected through its moving silhouette. In fact their bodies can be described as wafer thin. Furthermore it has developed a silvery body surface that reflects light so that it cannot be seen easily by predators looking at it from a horizontal direction. The most fascinating development of the Hatchetfish however is the appearance of photophores on their bellies. Photophores produce light in different colours. In the case of the Hatchetfish, the light produced matches exactly the light coming from the surface. Its predators are therefore unable to see the silhouette of the Hatchetfish, when looking upwards.

Unfortunately however for the Hatchetfish, some of its predators have developed very large yellow eyes, which enable them to distinguish light coming from the surface and that produced by the Hatchetfish photophores. The arms race unfortunately is as strong at this level as up on the earth's surface amongst humans.

Siphonophores are a colony made up from a large number of individual jellyfish that have linked together. They can be massive, many tens of metres in length. Each individual member of this colony has a specific function to ensure the continued existence and propulsion of the colony. Unlike colonies of bees and ants however, in the case of the Siphonophores, the individual members are actually linked together and have no freedom of movement outside the colony. Siphonophores have also managed to keep themselves nearly invisible like the original jellyfish.

-.-.-.-.-.-.-.-.-.-

The deep-sea floor has no light reaching it at all and is usually covered in mud.

One of the most astonishing creatures of the deep-sea floor is a bivalve mollusc that feeds off wood. It is assumed that it has developed the ability to eat the masses of wood that reached the seabed from sunken ships. The question that arises is what did they eat before man started using wood to build boats? The answer that it was fed on wood from trees carried by rivers may not provide a true presentation of facts, as at these deep levels of the ocean, you would hardly expect to find any trees carried by rivers. The only answer is that they developed this ability just during the last few thousand years, simply to take advantage of the food that suddenly became available. Darwinian randomness? Difficult to accept! It is more than likely that the Counter RNA once again provided a quick solution to the problem.

The most common creatures of the deep-sea floor are the sea cucumbers, most of which look like long, fat sausages. They crawl on the seabed sucking up the sediment to extract any organic materials. One of these sea cucumbers has managed to develop a webbed structure at the front and the back of its body. This web structure enables it to swim. It does actually venture as much as 1000 metre off the seabed, looking for alternative sources of food and to avoid its normal predators.

One of the worms of the seabed, that normally spends most of its time in burrows, the Biremas, has acquired a similar ability, having developed a tuft of tentacles around its mouth. When disturbed by predators it simply propels itself out of the vicinity.

Some echinoderms have found a permanent residence on rocky slopes near the seabed. They realised that this provides for them a good catching point for foods passing buy. To improve their chances off catching the passing food, they have developed a stalk and an umbrella of arms. In fact they look more like plants than animals. Sea lilies and stalked crinoids, which look like flowers, belong to this category.

Some rat-tail species have developed a light source near their anus. They then use a system of mirrors to direct the light forward to help them in their search for prey. This amazing feat does not end here. They have also developed a lens to ensure that the beam is concentrated in the right direction. Poor old Pythagoras! He thought he was the first to use lenses for a specific function! Unless of course these developments happened after Pythagoras viz. during the last 2,500 years! With the speed at which Counter RNA works you can never be too sure!

-.-.-.-.-.-.-.-.-.-

CHAPTER 7

REPTILES AND AMPHIBIANS

Let us now consider some examples from various species of reptiles and amphibians.

The sand swimmer snake moves under the sand. Its streamlined head has toughened glossy scales to cut through the sand. It has developed scratch resistant eyes and its nostrils have developed special valves to keep out the sand grains. Even the mouth is set back to reduce friction. Sensors on its skin can detect insect movements.

Two of these characteristics, the scratch resistant eyes and its nostrils with the valves that keep out the sand are absolutely essential to the survival of this snake if it were to live under the sand. Can you imagine this snake acquiring these two characteristics through simultaneous random mutations? Very doubtful indeed!

Only the Counter RNA could possibly introduce such characteristics.

Its other special characteristics even though not essential they are still obviously desirable for more efficient movement under the sand and one could possibly postulate that such mutations were again the work of the Counter RNA.

-.-.-.-.-.-.-.-.-.-

A species of frog that live in the Titicaca Lake at an altitude of around 4000 metres on the Andes in South

America has developed flaps and folds to increase the surface area of its skin, which in turn increases the rate of absorption of oxygen. This frog has a permeable skin through which it breathes oxygen from the lake. Due to the high altitude, the amount of oxygen in the lake is very low, and so the frog had to invent a method to ensure sufficient supply of oxygen, otherwise it could not survive.

In spite of this increase in its skin area, the frog sometimes experiences problems with the amount of oxygen intake. So it has invented another method to increase even further the amount of oxygen it breathes in. The frog has learned to do press-ups. This has the effect of increasing the amount of oxygen flow past it skin. As the oxygen flow past its skin increases the amount of oxygen intake also increases.

This is a very important example. If one was to accept random mutations (Darwinian evolution), this means that this frog would have to be created with all its special attributes at the same time, which more than likely would require a number of random mutations to take place simultaneously, otherwise it could not have survived.

Through my hypothesis however, this frog could evolve from another species (which already had the ability to survive under the extraneous conditions of low oxygen concentration), in a number of stages through a number of self-generated mutations that could take place over a long period of time.

-.-.-.-.-.-.-.-.-.-

A frog in Madagascar enlarges its size to scare a snake off and avoid being eaten by it. If this fails, the frog secretes a gummy fluid over its skin. If the snake tries to catch the frog, its jaws are temporarily glued together, by this gummy fluid.

Without this facility the frog would not survive. This is one of those situations that I originally thought that it must be a random mutation that created it. After all it is possible that a single mutation can produce this particular feature. However on reflection I had to discard this view. The main questions that random mutations cannot explain include:

Why only this particular frog that lives in an environment with numerous snakes has acquired this feature?

Why haven't cows, dogs, or any of the one million other animals acquired this feature?

For me, it is obvious that the frog must have exercised the pressure on itself to produce such a change.

-.-.-.-.-.-.-.-.-.-

The various stages of growth of the ordinary frog are a true miracle. Frogs normally lay their eggs in pools; the eggs hatch into tadpoles, usually during early spring. Tadpoles have gills through which they breathe and they also have a tail, which enables them to swim freely in water. They feed on algae and other vegetation. As the tadpole matures it goes through to the next stage by absorbing its tail, getting rid of its gills, developing lungs and finally growing legs. Then suddenly a frog jumps out onto land.

Some frogs however have developed some unorthodox ways of going through these processes. They all lead to the fact that their eggs are extremely vulnerable and some species have developed ways to overcome this weakness.

The rhinoderma is a very small frog that lives in Chile. It has developed an exceptionally peculiar way to look after its young ones. The female lays its eggs on humid ground and the male sits close to the eggs guarding them. At a certain time before the eggs hatch, the male picks the eggs up and appears to eat them. Instead of swallowing them however he deposits them into his vocal sac, where they are allowed to develop fully, until one day the fully developed frogs jump out of his mouth.

-.-.-.-.-.-.-.-.-.-

The male midwife toad carries the eggs laid by the female until they hatch. Unlike other toads the midwife toad lays its eggs on land. As soon as the eggs are released the male takes hold of the strands that carry 20 to 50 eggs and entangles them around his thighs. He then hides in a burrow, coming out at night-time to moisten the eggs. The eggs stay there for about a month. The male toad knows when the eggs are about to hatch. He goes into the pool and the tadpoles just swim away.

-.-.-.-.-.-.-.-.-.-

The burrowing frog, which is found in some very hot deserts, has developed short, powerful legs well suited for the purpose of digging. Some of these frogs have also sharp snouts and flat spade like toes that aid them in their digging. It burrows into the cooler soil of the desert to wait out the unpleasantness of the very high temperature outside. The burrowing frog can sometimes even hibernate in its burrow.

The question is: Did this frog suddenly, and by sheer chance, evolve such powerful legs and then decided to start burrowing into the soil to avoid the extreme heat of the desert? What about their sharp snouts or flattened spade like toes? Were they also developed through sheer chance?

Why is it that other frogs in cooler countries have not acquired these characteristics?

Surely the answer must be that these characteristics were not developed through chance but through necessity. This is what the burrowing frog required; this is what the burrowing frog got.

The water-holding frog Cyclorana lives in the desert of Central Australia, where on average rain appears once every two years. There have been instances where there was no rain in this desert for several years. As soon as the rain appears and the sand gets wet these frogs come out of their hiding place.

The rain normally produces a few pools of water amongst the hollows of the rocks and the frogs enjoy themselves thoroughly for a few days. During these few days the frogs survive on insects, which are also the outcome of the rains. They also mate immediately and they lay their eggs in the pools of water. Surprisingly enough the eggs hatch in a remarkably short time and the tadpoles develop at an amazing rate.

Before the pools dry up, the frogs including the young ones absorb as much water as they can through their skin to the extent that they become bloated and they appear like small balls. They then dig deep into the sand forming their own chambers. Somehow they secrete a membrane from their skin, which covers their body entirely except for some minute holes for breathing. This membrane stops the evaporation of water from their skin. They can live in this suspended animation until the next rains appear. Usually rain appears every two or three years but on occasions it might take quite a few years before rain appears in the Australian desert.

-.-.-.-.-.-.-.-.-.-

Many tree frogs have developed expanded adhesive discs at the ends of their toes, permitting them to climb up the smooth surfaces of tree trunks. Some members of an Asian family are even more specialized for arboreal (tree-dwelling) life. Their feet are webbed, with fan-like structures that enable the frogs to jump from one branch to another or from one tree to another. When they arrive at the next branch they use their adhesive discs on their toes to attach themselves onto a branch. They can also jump from a high branch all the way to the ground. They can actually jump from a great height over distances of many tens of metres by opening out their webbed feet and using them as parachutes. Even though they do not fly, they are sometimes called as flying frogs.

Two very important features for the tree frogs! Surely not through chance! Why haven't other frogs acquired these features, apart from those that live on trees?

-.-.-.-.-.-.-.-.-.-

The red-eyed tree frog has also developed special toe pads that allow it to crawl up the trunk of a tree and also to adhere to the underside of large leaves. The female selects leaves that overhung the water below and lays her eggs on the underside of these leaves. When the time comes for the eggs to hatch, the tadpoles fall straight into the water below. The eggs with so many predators in the water have a much higher chance of producing tadpoles when they are attached to the leaves above the water.

The comments made above for the "flying frog", can equally well be applied to the red-eyed frog.

-.-.-.-.-.-.-.-.-.-

Some snakes can spit their venom at distances in excess of two metres in a fine spray. They aim this venom at the eyes of the victim. The victims do sometimes include humans. The venom can cause blindness. The important thing to note here is the fact that the spitting of the venom is never used to acquire food, it is only used for defence, when the snake thinks that it is in danger.

The acquisition of the venom through a mutation is one thing, but the acquisition of the ability to spit it at a distance of over two metres is another. This requires a further mutation. What about the ability to aim? What about the acquisition of the knowledge that the venom in the eyes of its victims causes much more damage than anywhere else on the body of the victim?

Could these characteristics have possibly been developed through randomness? People who insist that such characteristics can be the result of randomness must obviously attach some divine properties to randomness.

Counter RNA surely presents a much simpler and much more beautiful solution!

-.-.-.-.-.-.-.-.-.-.-

The North American cotton snake has developed a number of Infra Red sensors that detect a mouse at night from the heat emitted by the mouse. Firstly it has a set of two heat sensors just below its eyes. As it approaches the mouse it opens its mouth to catch it. To ensure that the snake does not miss its target at that instant, the snake has also developed further sensors in its mouth, which help it aim with greater accuracy.

This is probably unique in the animal kingdom. To have eyes in the mouth for increased accuracy! How closer can one get to a perfect detection system to ensure that the snake does not miss its target? Surely this must have been designed by a process to ensure fulfilment of a need. Randomness could not have possibly been involved in such a design!

-.-.-.-.-.-.-.-.-

It is not unusual for many creatures to develop strong toxins in their skins to ensure that they do not form an attractive proposition to their predators. The poison arrow frog of South America is such a creature. In fact hunters used to wipe their arrows on the skins of these frogs to ensure a kill of a large animal.

Many species of newts have developed special glands in their skins that produce toxic secretions, which can be lethal to many of its potential predators. Some populations of garter snakes in western North America have however evolved the ability to feed on these newts.

This is the cat and mouse game that we have come across previously. However it would not have been possible to develop an antidote for the toxic substances emitted by the newts through randomness. These substances are absolutely lethal for much bigger predators than the garter snakes.

-.-.-.-.-.-.-.-.-

Snakes are extremely flexible creatures as their body can be made up by as many as 300 vertebrae. What is most amazing however about most snakes is that they can swallow much larger prey than the physical size of their mouths and even the size of their heads. They can achieve this because their skull is very loosely built and can be stretched in several directions. The two bones of the lower jaw are joined by a very flexible ligament and are attached to the skull by a short moveable bone. To reduce friction and make the passage easier, the prey is covered with saliva as it passes through the mouth.

My view on snakes is that after the original random mutation that created the species, any further development that created their special functions such as the shape of their mouths, was formed through the Counter RNA.

-.-.-.-.-.-.-.-.-.-.-

There is only one species of sea lizards and that is the Galapagos marine iguana. They feed on the algae that grow on the rocks close to the beaches. Because lizards are used to a warm environment, they find it difficult to venture into the cold seawater around the Galapagos Islands. They wait for the sun to warm them up before they go into the cold water to get to the rocks. The Galapagos iguanas have developed special claws on their feet that enable them to hold onto the rocks in spite of frequent strong waves, that could have otherwise dislodge them and throw them into the cold sea.

The larger male has however discovered that there are higher concentrations of algae in the deeper water around the rocks. They dive into the deeper water and manage to stay there for up to about 10 minutes, during which time they can enjoy the much richer pickings of algae. The smaller female is unable to do that.

To avoid getting chilled in the very cold water, the male iguanas withdraw their blood to the centre of their bodies delaying the cooling process.

Does this sound as if it was developed through pure chance? Or is it not more likely that the pressure of getting to the richer deposits of algae, the iguana managed to acquire this important characteristic through our Counter RNA?

And what about their special claws on their feet? They probably developed first, because without these claws the iguanas would really be in trouble. But again it is more likely that these claws were developed through the help of the Counter RNA.

-.-.-.-.-.-.-.-.-.-

The examples that follow are typical of developments that could only be achieved through the Counter RNA after the original random mutation to produce the species.

The horned toad can control its metabolic rate of production of heat in such a way that it has become a function of the heartbeat. As this is a reptile that lives in deserts it is highly desirable not to produce too much heat during the day when the ambient temperature is excessively high, whilst it needs a lot of body heat at nights when it can become quite cold. The horned toad achieves this by controlling the rate of its heartbeat.

Some tarantulas love frogs as part of their diet. However one such species of tarantulas cohabits with a species of frog in its burrow. This frog has developed a means of secreting a substance through its skin, which is distasteful to the tarantula. The frog also is safer in the tarantula's burrow, as some of its other predators are devoured by the tarantula. Furthermore the frog protects the tarantula's eggs from invading ants. Cohabitation with mutual benefits!

Some desert lizards find it difficult to stand in the hot desert sand continuously. They have developed a method through which they only touch the sand for half the time. They simply lift alternately two of their feet off the sand. They first lift up the left front foot together with the back right foot. After about 10 seconds of relief for these two feet, they bring them down and they immediately lift up the other two feet.

For the same reason, the sidewinder snake of South West Africa only touches the ground at two places at a time, to avoid being scorched as it moves on the very hot desert sand.

By utilizing the sun's heat as an energy source, the chameleon can survive in the desert on one insect for every 2-3 days. Similarly snakes in the desert, can survive on one tenth of the food required by mammals of similar body weight. Rattlesnakes only need twelve meals per year to survive. Many can go without food for months and in some cases up to a year.

A species of snake in Madagascar looks exactly like a dead tree twig. Its prey, mainly small lizards, completely unaware of its presence, walk straight into it.

Many chameleons can flatten their body to less than half their size in order to hide behind branches when they see a predator. When faced with the predator however they can breathe in air to double their apparent size to present a formidable opposition to the predator.

-.-.-.-.-.-.-.-.-.-

Many scientists subconsciously accept the fact that environmental pressure can produce changes to an organism. One can observe this in their writings or their speeches. The following extract from the Encarta Encyclopaedia demonstrates another important point. The author of this article clearly says, "This adaptation appears to be caused by environmental pressure". This author does therefore subconsciously support my hypothesis.

Axolotl, common name for the tadpole form of the yellow-spotted, brown salamander found in Mexico and the western United States. It is unusual in that it attains maturity and reproduces in the tadpole, or larval, stage of amphibian metamorphosis. Axolotls inhabiting Lakes Chalco and Xochimilco near Mexico City retain their gills, undeveloped legs, and finned tails and merely increase in length to about 25 to 30 cm (10 to 12 in). **This adaptation appears to be caused by environmental pressure**; as the surrounding country became too dry and barren to sustain amphibious animals, the lakes in which the axolotls are born provided cool, well-oxygenated water, good shelter, and an abundance of insect and small animal life for food. Until 1865 scientists considered the Mexican axolotl a distinct species, with no connection to salamanders. In that year, however, a number of Mexican axolotls on exhibition in an aquarium in Paris lost their gills and changed into salamanders. Later experiments proved that adding thyroid extracts to the water will induce or hasten the metamorphosis of Mexican salamanders kept in tanks.

This is obviously another excellent example where my hypothesis of the Counter RNA is fully justified.

-.-.-.-.-.-.-.-.-.-

Atrophy is an important aspect of evolution. Many reptiles have lost their legs in order to become serpents. This is evident on a number of snakes, including a species of python, where a pair of stumps appear in the place where the hind legs were.

The Australian blue tongue skink has only very small legs, probably on its way to losing them completely.

The Australian scaly foot has only a pair of stamps to show that it is a burrowing lizard. Obviously if lizards have taken to burrowing, they have very little use of legs. In fact they can be a hindrance.

The Counter RNA comes into service once again, to help remove unwanted parts, to make life more efficient and comfortable.

-.-.-.-.-.-.-.-.-.-

The alligator snapper turtle lives in the Amazon basin and lives virtually entirely on fish. It stays still whilst waiting for the fish to come to it. It fishes with its enormous mouth open. Inside its mouth is a bright red filament that looks very much like a worm. When a fish sees this worm it goes for it. All that the turtle has to do is to close its mouth and it has a meal. To ensure that fish notice the bait, it frequently makes it move in such a way that fish think it is an actual worm moving.

-.-.-.-.-.-.-.-.-.-

Red crabs are the most distinguishing feature of Christmas Islands in the India Ocean. There are over 120 million red crabs on a small island with an area of only 135 square kilometers. Red crabs live amongst the dense vegetation of the island eating mainly dry leaves. In September every year they make a burrow into which they spend the dry months of the year up until November. They always cover the burrow with a large dry leaf.

As soon as the rains arrive around the middle of November, the 120 million crabs come out in unison and they all start a long walk to the beach, which can take a few days. It is estimated that around a million of them are killed, on this arduous trip, which takes them over roads as well as high dangerous cliffs. When they get to the sea they seem to be rejuvenated. They stay there only for a short period of time and then they start on the return journey.

As soon as they go back to the dense vegetation the males start digging the dens into which the females will go into after mating. There are ferocious fights during this period between males. The females go round selecting their preferred male that has completed a den for them. There are no prenuptial contact procedures between male and female and the actual mating lasts for just a few minutes.

The female immediately goes into the den, where she stays for about two weeks. After this period, all the females come out of their dens simultaneously and they all start making their way to the beach.

Once more there are many killed on the way to the beach, but this is something that they have to do. When they get to the beach they get into the water and start a peculiar dance whilst standing up. This dance enables the dislodging of the eggs from their bodies. The eggs simply fall into the seawater where they are instantly hatched and tiny crabs appear.

Every crab produces around 100,000 eggs. Most of these millions of young crabs are eaten by predators or carried away by sea currents. But millions of them survive. One month later the young crabs get off the sea and start making their way into the dense vegetation of the island to get together with all the other crabs. It is a mystery as to what makes them get out of the sea and how they find their way to exactly the correct spot.

-.-.-.-.-.-.-.-.-.-

CHAPTER 8

BIRDS

Let us now consider some examples from various species of birds.

Migratory birds use various mechanisms to find their way whilst travelling thousands of kilometres. Guides include the magnetic lines, the moon, the sun, the stars, smell, topographic landmarks, etc. If they are to use the sun as a guide they must use some process by which they compensate for the position of the sun at any part of the day, modulated of course by the time of year. This is a phenomenal achievement, for they must have mechanisms to sense all these and take appropriate actions. Apart from the sensing element they must have some facility of storing information on landmarks as well as information on the position of the sun based on the annual calendar. It is difficult to see how they can store such an unusual amount of information in their tiny and primitive brain, but through one way or another they seem to do it.

If one navigational system fails, they revert to a backup. Unfortunately these guides fail them frequently. There is no sun to guide them at night time or during dark days. The moon is not always available either. Magnetic lines can be interfered with especially during magnetic storms on the sun. The stars are not visible if the sky is cloudy. The topographical information available to them is useless at night time.

Thus the birds, as well as other migratory creatures such as butterflies, can lose their way. Also artificial lights such as lighthouses can become a source of disaster if the birds come across them at night, since they try to use the lighthouses as a guide and frequently they fly straight into them.

Migratory birds usually travel together in big swarms but sometimes they fly alone. They usually fly away from temperate and arctic zones to warmer climates, at the end of the summer.

The incredible thing about some migratory birds such as the house martin is that newborn offsprings can travel alone all the way to Africa. Also when they return, they have no problem locating the same nest that they had the previous year. This is more amazing considering the fact that they travel a few thousand miles, many of which are over the Sahara with virtually no landmarks or over the Mediterranean Sea, where there can be no topographical reference.

Obviously the route must be mapped genetically on the martin's brain. How else can this behaviour be explained? But Darwinians say that experiences of a lifetime cannot be genetic. This is another matter of concern for Darwinian evolution.

-.-.-.-.-.-.-.-.-.-

One of the best examples cited by Darwin concerning birds is the example of finch birds on the Galapagos Islands. He explained that during the last few thousand years, some small birds from the mainland were blown by the wind to the Galapagos Islands. Within a short period of time a number of species evolved to take advantage of the local resources. The main features concerning these new species were the shape and size of their beaks. The species that have adapted to seed eating have large, heavy beaks. The insect eaters have small sharp beaks.

-.-.-.-.-.-.-.-.-.-

The Grand Master however in his first book, the Origin of Species, when discussing the woodpecker, makes the following comment:

"But in the case of an island, or of a country into which new forms could not freely enter, places in the economy of nature would assuredly be better filled up if some of the original inhabitants were in some manner modified".

He continues to say that on some islands where woodpeckers did not exist, some other animals appeared to fill the gap of the woodpecker. The woodpecker uses its strong beak to break the bark of trees making holes to access insects that lie underneath the bark. There are however a number of alternatives that have developed in such islands such as the akiapolaan in Hawaii, the woodpecker finch on the Galapagos, etc.

I frankly consider Darwin's above statement and the examples that he gives as truly supporting my hypothesis all the way. What he is talking about is very similar to my statement of pressure to evolve. In this instance he is talking in terms of pressure of availability of a resource that some species could take advantage of.

This has nothing to do with randomness. Darwinian evolution assumes that a certain species such as the ordinary finch, acquires through randomness a characteristic whereby it can make holes on the bark of trees to get at insects that exist there, simply because woodpeckers were not there to get to them first. But the question that arises is "why didn't these species appear through randomness in other parts of the world"? Admittedly in the presence of the woodpecker there would have been strong competition and the woodpecker would have probably won it, but why is it that we have never found any traces of creatures that ever existed that competed with the woodpecker? Simply because such creatures never existed.

-.-.-.-.-.-.-.-.-.-

Dippers survived for millions of years as a normal bird when it suddenly acquired the extraordinary ability to dive into freezing waters for its food. They manage to walk and swim (using their wings) underwater in search of their food.

The dipper has developed oily wings and a thick layer of down which keep it dry and warm whilst feeding in the cold water of a river bed or of a lake. It has developed the ability to store enough Oxygen to survive under water for many minutes. It has also developed long toes and very strong claws, which allows the dipper to grip firmly on the rocks on the bed of a fast flowing river without being swept away. It can even withstand the force of whirlpools and ferocious rapids. It has the habit of slightly tilting its wings against the fast flowing water whilst walking at the bottom of rivers and streams searching for its food. Their food consists of fish, tadpoles, fish eggs, larvae, etc.

A number of absolutely necessary features for the dipper to survive the way it has chosen to search for its food. There is no way that the dipper could survive in its way of life without possessing all the characteristics described above. It is difficult to see how random mutations could provide the dipper even sequentially all these features.

-.-.-.-.-.-.-.-.-.-

Eagles fly higher than any other bird. The eagle has developed an eyesight that is probably unique in terms of its ability to focus at a great distance. It can spot a small hare or a rat from a distance of 3 kilometres.

One could ask why couldn't a random mutation create such a characteristic? I agree! Except that this eyesight was not developed by the rat, the rabbit, the hare, or any other prey of the eagle. They had no need for it. It was developed by and for the predator that actually needed this feature.

-.-.-.-.-.-.-.-.-.-.-

The swordbill hummingbird has developed an extremely long beak, which in some cases it is up to four times longer than the rest of its body. They have also acquired a long extensible tongue. The hummingbird needs this long beak to collect nectar from flowers, which are conical in shape, and their nectar is found a great distance from the top of the flower. To be able to achieve this, the hummingbird has developed the technique of hovering on top of the flower and pushing its beak into it. This is a technique that very few birds have acquired even to a limited extent. Each wing of the hummingbird is jointed so that it moves in a figure of 8, so that it creates a lift on both the backward and forward strokes.

When faced with larger flowers it actually flies into the flower and hovers within the flower. After it picks up the nectar, the hummingbird flies backwards, to get out of the conical flower. It is the only bird that can actually fly backwards.

When one examines all these features and bearing in mind that this bird acquired its unique features only because of the particular flowers in its environment, one hesitates to accept that such features were developed through randomness.

-.-.-.-.-.-.-.-.-.-.-

The knot, a species of wading bird on the English coasts can detect the fact that the seawater has gone out in a tide whilst they are a few kilometres away and without having a view of the coast. These wading birds, which live in large flocks, fly away as the tide rises and go to nearby lakes or other areas where they can stay until the seawater recedes. They stay there waiting. As soon as the water recedes immediately they fly towards the sea. They are the only bird that can sense the position of the tide from a long distance.

How did the knots acquire this ability? Has the knot developed a timer modulated by a calendar? If that is the case then this is a fantastic characteristic that can only be explained on a principle similar to my proposal for the Counter RNA.

It is however possible that the knot may be using the properties of some unknown force of nature that we have not as yet detected and of which we have no knowledge. I sincerely believe that this should be investigated thoroughly to ascertain such a possibility.

-.-.-.-.-.-.-.-.-.-

The great grey owl of Canada has developed special hearing aids that can be described as the ultimate amplifier for stereophonic sound. This hearing aid can actually filter out any sounds outside the frequency that the owl is tuning into. From a distance of more than ten metres it tunes into the high frequency emitted by a lemming whilst moving under a thick layer of snow. The owl can then pin point the position of the lemming through its stereophonic detection system and flies directly to it. It can detect this sound even in the presence of wind or during a snowfall. Before beginning the assault, the owl uses both his hearing as well as his vision to focus at the precise position of the lemming under the snow. Even if the lemming starts moving, the owl uses a mechanism by which it compensates for the movement of the lemming.

To ensure that the lemming is not aware of the imminent attack, the owl has also developed special soft velvety feathers that ensure a virtually silent flight.

This owl, which lives in the northern latitudes where it experiences the midnight sun, it has no option but to hunt during the day, unlike other owls which are purely nocturnal. It has very few options in terms of choice of food. That is why the lemming provides a life saving threat to the owl. It simply had to develop ways of catching the lemming.

The question arises: What are the chances of the owl developing both these highly desirable features by sheer chance?

-.-.-.-.-.-.-.-.-.-

Albatrosses are nomadic birds that can stay in the oceans for many months. They can actually stay up on the air flying for periods up to about 12 hours. They can fly and look for food below day and night. They usually sleep on the sea. These flying machines can travel thousands of kilometres at every sortie looking for food. It is not unusual for albatrosses that are normally around South Georgia, to venture up to the coast of Brazil, many thousands of kilometres away, travelling at up to 90 kilometres per hour.

They achieve all this through their magnificent wings, which have been specially developed to have a span of up to about three metres, to be very thin and light in weight, which enables them to glide for up to 90% of the time that they are up in the air. Not too much chance of randomness, to achieve such virtual perfection. Without such perfect wings they just wouldn't stand a chance.

-.-.-.-.-.-.-.-.-.-.

The examples that follow are typical illustrations that give support to my hypothesis.

The most characteristic aspect of birds is their feathers. Feathers form one of the best heat insulators available. This is extremely important for keeping birds warm. They have also developed the best aerofoil structure and comparing their weight they are better than any aerofoil that man has ever achieved. This is because their microscopic structure is composed of millions of minute filaments, which are interweaved so that they produce maximum tensile strength, high aerofoil properties and through the use of air bubbles produce a very effective insulator.

Bird feathers are such effective heat insulators that penguins can survive in the cold winter of the Antarctic, the coldest place on earth. Their feathers are made of a number of filaments which trap air in between them forming an excellent heat insulator. Penguins have in addition to their feathers developed also a thick layer of fat under their skin. No other creature can survive the conditions of the Antarctic, which reach 40 degrees below zero.

Guillemots are very good divers and swimmers and feed mainly on fish. They dive down to the bed of the sea looking for prey. Some of them live on small islands and they sometimes fly over great distances over the Atlantic Ocean. In situations where sudden strong winds develop over the ocean, the guillemot takes a dive from the heights that it may be flying and starts flying as close to the sea surface as possible, or even goes into the sea itself. This is because the guillemot realises that if it stays high up it may get blown away deeper into the ocean and will not have the strength to return.

The kingfisher can pin point a fish from a distance of over seven metres. This is obviously of great advantage for those birds that live almost exclusively on fish that they catch. It also has the facility of hovering above water to look for prey.

The vulture lammergeier picks up the bigger bones that it has already stripped of any meat and carries them to a very high altitude from which it drops them on hard rock. The bones break revealing the bone marrow on which it feeds with obvious satisfaction. This is the equivalent of using a tool such as the hammer and anvil technique used by the chimpanzees to break palm nuts.

Most birds that dive under water for their prey have developed oily feathers that keep the water away so that they can fly as soon as they get out of the water. Darters and cormorants however need to go deeper for their prey and they cannot last too long under water before they run out of oxygen. So their feathers have developed differently. They absorb water, which makes them heavier, thus reducing the buoyancy, which makes it easy for them to dive at greater depths and at a much higher speed. When they finish fishing, they simply have to stand on a hard surface for a few minutes and open their wings to dry them, before they can fly.

The pink colour of flamingos is due to the small aquatic crustaceans that it feeds. They catch these by pumping big quantities of water together with the crustaceans, and then by pushing the water out through a filter device in their beaks. This is a similar technique adapted by the blue whale.

Ducks of various species do normally congregate together in thousands during the spring. The drakes usually develop some features on their heads of different shapes and colours that are characteristic of each species. This is necessary so that the ducks of each species can single out their male partners. When one of these species find themselves on small islands they no longer need to have these features to separate them out, as they are the only species on the island. In a very small period of time, they stop growing these colourful features.

The arrival of birds on small islands can have another effect on them. For some unknown reason, when a bird reaches an island which has no four footed predators, it becomes flightless. One possibility is the fact that flying for many birds involves a lot of energy. So if a bird realises that such energy expenditure can be avoided, by simply not flying, it simply stops flying. There are examples of these on the islands of the Great Barrier Reef in Australia as well as the Galapagos, Aldabra, and New Caledonia.

The male bowerbirds have developed a strange way to entice their female partners. The bowerbirds construct a beautiful shelter or bower, which they decorate with any items that they can find and are of a particular colour that they like. It picks up twigs, which it arranges in a beautiful formation to form the shelter and then starts bringing the items with which it decorates the shelter. It brings flowers, leaves, seeds, fruits, feathers, sticks, stones and any items that it finds as long as it likes the colour. He has two entrances, a front and a rear. The preferred colours are yellow-green and blue. One species extracts a dye from blue berries, which it uses to paint the twigs; he uses a fibre to do the painting with. Another extraordinary tool user that has been around for rather a long time! He then sits there and waits for an interested female to turn up. When one does, he immediately gets very excited and picks the items with his beak and shows them to her! If the female is suitably impressed they immediately proceed with mating, which takes place either within the shelter or just outside it.

The penguins have a problem in being unable to distinguish male from female; they are indistinguishable even to their own kind. The male has therefore developed a technique when it is looking for a female to mate. It picks up a pebble with his bill and lays it in front of another penguin that he has singled out. If that is another male penguin it responds by squaring up to him ready to start a fight. If it is a female that has already mated or not ready to breed, then his offer is treated with disregard; the female simply walks away. When he makes the right choice the selected bird bows downwards and picks up the given gift. He immediately bows downwards as well, as a gesture of happiness that he has found the right mate; they both open out their wings and start embracing each other.

The cuckoo is probably the best well-known example of a brood parasite. It lays its eggs in the nests of other birds. As most of the other birds that the cuckoo chooses to lay its eggs, have small eggs, the cuckoo's eggs are also small. Other birds of similar size as the cuckoo have bigger eggs. To ensure that the other birds do not detect the presence of its eggs, it ensures that the colour of its eggs matches that of the eggs of the bird in whose nest it lays its eggs. It lays its 15-20 eggs singly in various nests. The birds they choose to nurture their young ones are always insectivorous as cuckoos are insectivorous themselves. To increase its chances of survival, after the young cuckoo hatches, it usually pushes some of the other young birds out of their own nest. It actually pushes out any eggs that have not yet hatched as well. It does this with the use of an extraordinary hollow that has been created on its back. The young cuckoo puts its head under the egg and pushes it upwards over its head into the hollow on its back. When the egg is in the hollow of its back, the cuckoo lifts itself up and throws the egg over the nest.

In order to keep their eggs away from predators, some birds lay their eggs on ledges on high sea cliffs. To ensure that these eggs do not roll off, their shape has been developed in such a form that they are pointed at one end; this ensures that if they roll, they simply roll in a small circle and not down the cliff.

The Australian mallee fowl has developed its own incubator, which allows it freedom whilst the eggs are being incubated. This incubator, which is built by the male bird, consists of a quantity of rotting vegetation, which is then covered by sand. During the incubation period, which lasts over five months, the male bird sits outside and does regular tests on the temperature inside. He does this by penetrating his bill inside. As the outside temperature rises, the rate of the decaying process increases and therefore the temperature inside also goes up. He immediately removes some of the covering sand to allow heat to escape and thus keep a constant temperature inside. When the sun is shining directly on the mound, he piles more sand thus creating more heat insulation to ensure that the heat inside remains constant. Later in the year as the outside temperature drops, he opens up the top of the mound to allow the sun to warm it up during the day, and then closes it again at the end of the day to ensure that the heat stays inside the mound.

The hoatzin bird of South America is probably the most strange looking bird in the whole world. It has got beautiful plumage, but many people think that its shape is similar to that of a reptile. The young chick of the hoatzin has claws, which later on disappear. However these claws are quite useful to the young hoatzin as it has a number of predators. As their nests are usually over water, the young hoatzin jumps into the water to escape its predators. It then swims back and using its claws it climbs back to its nest.

The bald vulture has a habit of cleaning itself soon after a meal. It likes to clean itself regularly by washing in water in a lake or even a small pool of water. It also uses its beak to pick up any small pieces of meat that may have got stuck on its feathers whilst enjoying its meal. There is one problem however with the way it cleans itself. It is very difficult to clean its head. The head of vultures usually gets covered in small pieces of flesh, because vultures push their heads deep into the carcass to scrape off as much flesh as possible. The bald vulture's head however does not get covered in flesh, because it has no hair. So it does not need any cleaning.

Most birds clean themselves regularly. After all getting rid of accumulated dust makes them lighter, so they don't need as much energy to fly. They can however be infected by fleas and lice which can cause big problems to the birds if they remain unchecked. So they allow insects to get into their feathers to do a clearing job on the fleas and lice. Some birds visit ants' nests and sit on the nests so that ants get into their feathers. Some birds actually pick up ants carefully with their beaks and push them into their feathers. Sometimes they squeeze particular species of ants that emit formic acid onto their skins to ensure that the parasites are killed by the formic acid. It appears that some birds also get pleasure out of scratching their skins, but it is quite possible that this is due to itchiness. It is not uncommon to see them on the ground lying upside down and moving their whole body quickly in such a way as if trying to pick up dust from the ground.

-.-.-.-.-.-.-.-.-.-

Kittiwakes are birds that build their nests on high cliffs. They build their nests by plastering together mud and water plants that they pick up from the sea, and on which they trample with their feet in order to make them tight. They form a hollow in which they lay their eggs. Whilst going through the motions of building their nests, due to the high numbers of kittiwakes that sometimes congregate together, there are continuous territorial disputes, which are always settled through prolonged duels. Kittiwakes are extremely aggressive and they do not give up easily. They fight with their beaks like swordsmen used to fight their own duels with swords. However the kittiwakes are extremely gallant. As soon as one of them accepts defeat the fight immediately stops. No retributions, no reprisals! How do they accept defeat? Simply! The one that feels that has lost the duel hides his "sword" under his wing!

The eye of the owl is basically a huge lens that gathers light. As the owl is a nocturnal bird, colour is not of much use to it. So the number of cones on the owl's retina for colour detection is very small. On the contrary the number of rods, which are sensitive to light, is very high. To increase the amount of light entering his eye even further, the eye has been elongated to accommodate a bigger cornea as well as a bigger lens. The position of the eyes of the owl also enables it to have a very good perception of distance. To improve its chances of catching prey, the owl has developed a very silent flight. However the most important aspect of its success is its hearing ability and the fact that it can locate its prey extremely accurately through the noise that the prey is making. The owl has achieved this accurate location ability in the non-horizontal plane, due to the fact that its ears are not symmetrically situated on its head.

-.-.-.-.-.-.-.-.-.-.-

The African sand grouse sometimes nests many kilometres away from standing water, which makes it difficult to ferry water to its chicks. The male grouse has however developed a unique feature to achieve this. After it has drunk enough water for itself, it creeps into the water allowing the feathers of his belly to get fully soaked. These feathers can absorb water like a sponge. When it goes back to its chicks, they sip the water from the wet feathers just like young mammals drink their milk off the nipples of their mother.

-.-.-.-.-.-.-.-.-.-

CHAPTER 9

INSECTS

Let us now consider some examples from various species of Insects.

Flowers appeared on earth 100 million years ago. They developed nectar to reward insects visiting them in order to transfer pollen. They also developed colours and scent to attract them.

Some flower colours can only be seen by insects in the UV region.

Bees (insects) and flowers have to coexist and develop together.

-.-.-.-.-.-.-.-.-.-

The Bolas spider has developed a unique way of catching its prey. It releases a thread with a sticky droplet (bolas) at the end of it. Some of these spiders swing the string to increase the chances of catching a moth. Other spiders throw the bolas as a moth approaches. The thread with the bolas at the end of it is attached to one leg of the spider. To attract the male moths to it, the spider releases a deceptive pheromone that is closely similar to the sexual perfumes of female moths.

It even modifies the released type of pheromone to attract other moth species at different times of the night as some moths come out in the early evening and others come out later on at night.

How could random mutations develop such features is something that only Darwinians could apparently comprehend.

To develop a sticky substance is one thing. That I would accept could happen through a random mutation – thank God it didn't happen to house pets and farm animals. But then to produce a suspending thread and attach the sticky substance at the end of it, through another random mutation, well that is really taxing randomness. In addition to these, to learn to swing this thread with such methodology and planning must really show some "divine" randomness. And then it also creates a female sexual pheromone. As if that was not enough it also mimics the sexual pheromone of a number of moths. But that is not the end of the inventiveness of this apparent randomness! The bolas moth knows what pheromone to release at what time to attract the right moths! Come on ladies and gentlemen! Are we serious about what we mean by randomness?

-.-.-.-.-.-.-.-.-.-

Insects developed wings 380 million years ago. Dragonflies are extremely efficient flying machines. In spite of their small size, they can fly at speeds of up to 30km/hr. They swing their wings at around 150 strokes per second. The two forewings move in opposite directions to the hind wings. When the forewings lift up, its hind wings push down. Their wings are designed in such a way that they get a lift on the downbeat of the wing by twisting it so that the leading edges incline downwards. At the bottom of the stroke the wing is twisted back so that it gets a lift on the upstroke as well.

It sounds like a group of very competent and very experienced aeronautical engineers must have been involved in the design of this equipment. In fact today's aeronautical engineers have not been able to match this technology.

Does it sound to you as if this can be the work of a large number of random mutations? Or do you think that Counter DNA provided a helping hand to the dragonfly during its efforts to improve its flying abilities?

-.-.-.-.-.-.-.-.-.-

A mantis has virtually the same appearance as the pedals of the ginger flower. This mantis sits motionless at a suitable position on the branches of the plant and pretends to be one of the flowers of the ginger plant. The mimicry is such that butterflies attracted to the flower, do not realise that this is not a true flower and are caught by the mantis.

If this mantis had developed this shape in isolation away from the tropics where this ginger plant grows, I would have accepted that a random mutation could by sheer coincidence produce that shape of mantis that looked like a ginger flower. But it didn't! This mantis evolved at precisely the right geographical spot to take advantage of the shape of the ginger flower. I therefore have to discount the possibility of random mutations and assume that once more the Counter RNA produced the magic wand to provide the solution that the mantis was looking for.

-.-.-.-.-.-.-.-.-.-

Honeybees normally live in hives of limited volume. They have therefore developed smaller but faster wings – due to limited space in the hive and the size of flowers they visit. One of the most incredible things about the honeybees' hive is the shape of its honeycomb cells. These are hexagonal with exactly 120 degrees between each one of the sides. This can be proved mathematically that it is the optimum shape for holding the maximum amount of honey and also for maximum structural strength relative to the amount of wax used. It is a truly remarkable achievement, especially bearing in mind that this is accomplished instinctively by groups of bees without any supervision and always working in the dark.

-.-.-.-.-.-.-.-.-.-

Ordinary flies have few defensive mechanisms against their predators. Bees however have their sting, which they use if they are attacked. A number of species of the Bee Fly have developed their bodies to look like bees – an example of protective mimicry. Their colour and growth of hair is also such that they look very similar to bees. Different species of Bee flies have developed so that they resemble different species of bees.

Is this a random mutation at work? Let us assume that one species of bee through a random mutation changed its body size and shape to look like a bee. Nothing wrong with that! But then come the colours. How did the right colours develop? The possibilities are endless. There are hundreds or thousands of colour combinations that the Bee fly could have acquired. But no! It acquired exactly the right colours to look like the bee whose body it mimicked. What are the mathematical chances of this occurring through randomness? Then there is the hair at exactly the right places of the body! Who do they think they are kidding?

And finally the fact that a number of species of Bee flies have developed in such way so that each one looks like a different species of bee! And all this through randomness!! Come on!

-.-.-.-.-.-.-.-.-.-

Leaf cutter ants live in colonies of many millions of ants. The worker ants slice off pieces of leaves using their sharp jaws. They carry these pieces of leaves back to their nest. Usually the size of these pieces is more than double the size of the ant. However the ants cannot digest the cellulose of the leaves. They chew the leaves and they leave them there for bacteria to feed on the chewed leaves. A fungus is the product of the bacteria interaction. This fungus can then be eaten by the ants. They can easily digest the fungus.

These ants which have been around for millions of years must have obviously been the first farmers, much before Cro-Magnon man ever thought of farming.

Was the development of this technology the result of random mutations? More than likely the ants started eating the fungus from naturally rotting leaves before they realized that they could produce this fungus in their own nests much more effectively. This process of learning (which has to be genetic) must have reached their DNA through some means. Again there is a high likelihood that the Counter DNA played its role in this DNA coding.

-.-.-.-.-.-.-.-.-.-

Amazon ants frequently invade the nests of other ants and carry away the pupae of the invaded colonies. When they invade other nests they find resistance from the soldiers of the invaded nest. However the Amazon ants emit a chemical that makes the defending soldiers flee away form their nests. They can then carry away the pupae to their own nest in peace.

When the young brood of the invaded colony grows up they are turned into slaves by the Amazon ants. The slaves tend to all the needs of the Amazon ants. They even feed their masters. In fact the Amazon ants have got so used to be tended by their slaves that in one laboratory experiment where a number of Amazon ants were left on their own with food next to them they nearly died of starvation. When one slave ant was brought to them immediately it started feeding them, and they managed to survive.

What made the Amazon ants to depend entirely on their slaves or was it something that they acquired through many generations of dependency upon their slaves? Could this be a random mutation? An acquired dependency on ones slaves? Surely not!

-.-.-.-.-.-.-.-.-.-.-

One other species of ants, the army ants, that normally live in colonies of about 250,000 behave in an admirable way whilst looking for their daily food. All the workers of the whole colony come out of their habitat early in the morning, and start an assault against any type of life that happens to be on their way. With such a big number of ants, nothing stands a chance. All larger animals run away on the face of this attacking army of ants. Smaller animals, lizards, insects, leaves, etc are quickly shred to pieces by the sharp teeth of these ants and quickly carried back to their nests. They are one of the most feared organisms of the animal kingdom and for good reason. It is estimated that they make over 50,000 killings every single day and that they completely destroy thousands of square meters of vegetation.

All this operation is carried out under the close look of the soldier ants that are much bigger than the workers and keep guard on either side of this phenomenal procession of the worker ants. There is no leader. They all work collectively as one group.

Many people believe that these ants use the position of the sun as a guide concerning the direction in which they operate. The incredible thing about their behaviour is that every morning when they come out of their nests they start their procession in a direction that is 123 degrees clockwise from the direction they followed the previous day.

One wonders as to why they have selected this strange angle of 123 degrees to start their new foraging sector every day.

Let us assume that instead of 123 they had selected 120 degrees. Obviously under these conditions, after 3 days time they would be back to the same sector that they foraged before, and there wouldn't be too much there, only 3 days after the previous decimation. So their pattern of moving forward 123 degrees (rather than 120 degrees) ensures that after 3 days they find themselves 9 degrees forward compared to the first position 3 days earlier.

This is a fantastic achievement.

Now let us consider some further implications of this particular angle. Why not 122 degrees for example? The reason behind this is that if they had moved forward by 122 degrees, in 3 days time they would be only 6 degrees ahead of the starting sector. However it appears as if the foraging sector they cover in each outing is more than 6 degrees wide. It is apparently close to 9 degrees. So it is essential for them that they cover 123 degrees as a minimum, since the foraging sector is about 9 degrees.

The incredible thing is that they know this in advance. Their ability of foresight and planning is admirable. However this is not the end of their foresight and planning.

If the ants continue at the same rate viz. 123 degrees rotation every day, it will take them 40 days to completely cover the whole patch of land around them. In other words they would completely decimate everything in that area after 40 days. The ants however suddenly after exactly 20 days move forward to a new location, carrying all their pupae with them.

The question that arises is why do they move on whilst only about half the area has been eaten away? Why don't they tackle the other half?

The answer may be simple. Is it possible that these ants do not destroy the whole area so as to allow the other half to be used as a centre of new life to start again in that region? In other words they want to give it a chance to recover under the best possible conditions. For if they turned the whole area into a desert, and they continued doing this in other areas as well, there would be no future for the ants themselves.

They travel forward for 10 days before they find a new suitable temporary nest. There is no such thing as a permanent nest for these ants.

Not only mathematicians that invented the number of degrees in a circle, million of years before Pythagoras, but also short term, medium term and long term planners with an intuition and foresight to compete with any other organism on this planet!

-.-.-.-.-.-.-.-.-.-

The fortuna female firefly has developed a system through which it mimics the sexual female signals of other species of smaller fireflies. Males of the other species are lured to these signals expecting to find a female of their own species. Instead however, the female fortuna turns to its male suitors and makes a meal of them.

How could this firefly find the way to mimic these signals of other species? Of Course Darwinians will insist that this was purely a random mutation. How convenient!

-.-.-.-.-.-.-.-.-.-

Most insects consume their food as soon as it is available. Ants, not only store food for later use, but also are well known for their farseeing concerning the distribution of the stored food. When they collect seeds, which they store carefully, to ensure that they are kept dry, they know well in advance which seeds to use when. Their offspring have different nutritional requirements at different stages as they mature, so they are being fed from the available stocks accordingly.

-.-.-.-.-.-.-.-.-.-.-

The dartling desert beetle covers all its water needs by drinking the water that condenses on its body from the dawn mist. It has very long legs and in trying to reduce the time that its feet touch the scorching desert sand it has become the fastest beetle in the world.

-.-.-.-.-.-.-.-.-.-

The Mono Lake in California is a lake whose water is so salty that no ordinary creatures can live in it. Towers of limestone surround it. The only life in this desert is primitive algae and millions of brine flies that feed on these algae. These flies have no predators and they have no competitors for the algae either. At the peak of the season their total weight can be as much 2,000 tons.

Their larvae also feed on the algae, so that the flies have to search for algae in the salty waters. They achieve this by submerging in a bubble of air!

-.-.-.-.-.-.-.-.-.-

A cricket that lives at a very high altitude in New Zealand manages to survive the freezing conditions that exist at night. Normally animals cannot survive the condition where the water in their body forms into ice crystals. As soon as this happens the cells of their bodies rupture causing permanent damage that leads to the death of the animal.

This cricket however has developed some chemicals in its body, which slow down the freezing process in a way that the cells adapt to the changing conditions and they don't get damaged. The following morning when normal temperatures return the cricket thaws and returns to its daily activities without any consequences due to the previous night's experiences.

-.-.-.-.-.-.-.-.-.-

The material, from which the exoskeleton of many insects is made, can have the tensile strength of steel. The material, chitin, is similar in structure to ordinary fibreglass.

Similarly the strength of the silk used by the tent moths for making their cocoons is demonstrated by the fact that a single thread can easily hold the weight of the moth. If these moths fall accidentally from their tents, they spin out a lifeline from the silk that they use to make their cocoons.

To reduce the danger from their predators many butterflies have developed patterns on their wings, which makes these butterflies appear similar to the predators of their predators. Thus for example, it is very common for butterflies to have patterns on their wings that when open out, give the impression of big eyes such as those of a large owl. These patterns are on the topside of their wings so that they become visible when flying or whilst sitting on a flower sipping nectar. This scares many predators away.

The female noctuid moth exerts an abdominal scent that the male can detect from several kilometers. Male moths and flies have developed antennae that are covered by millions of scent receptors, which can detect female aphrodisiacs.

The alder fly lays up to 500 eggs on a grass stem growing in a pool of water. It does this because the chances of survival outside the water are greater than within. To ensure however that the eggs remain dry in case that the water level rises, the eggs are surrounded by a waterproof transparent shield. One end of the egg needs to have access to the outside world, in order to breathe. Access is provided through a tiny pore on the shield. This part of the egg can get wet without any harm to the egg.

The colour of many insects that live in deserts has evolved to ensure two main purposes. The first of these is to ensure a camouflage against the background in which it normally lives. This is obviously a natural defensive mechanism against their predators. The second is to ensure maximum reflection of the sun's light so that they do not get overheated. The black colour associated with some desert insects has been given the explanation that these insects need to warm up as soon as possible from the sun early in the morning, following the very low temperatures that exist in the deserts during night time.

Most insects have a developed a phenomenal speed of response for their sense of vision. Thus for example humans have a maximum ability of resolving individual images presented to them one after another of about 50 per second. In the case of cinema where individual frames are presented sequentially, the eye is unable to detect the fact that only 16 frames per second are used. Most insects can resolve up to about 300 images per second. This is because they need this high resolution if they are to see clearly below them as they are moving at high speed above the ground. This high resolution is also necessary when they are chasing their prey, and the prey is changing directions very swiftly.

-.-.-.-.-.-.-.-.-.-

Plants always try to develop mechanisms to keep away animals that eat their leaves. Many species of milkweed are among the most dangerous of plants that have developed the ability to produce poisonous chemicals. The poison from the foliage of milkweeds contains a complex resinous compound that produces acute muscle-spasm seizures, symptoms of intense depression, and weakness in animals that eat them. It can even give them a heart attack.

The caterpillars of the monarch butterfly however eat the leaves of the milkweed without any effect on them whatsoever. These creatures have developed a mechanism through which they can separate the toxic substances in the leaves. They then transfer these toxic substances and store them safely in their own bodies, so that they then become toxic to their own predators. The yellow bands on the back of the caterpillars are a warning to potential predators of the fact that they are toxic.

To make things more interesting, the butterfly that evolves from the caterpillar takes with it this toxicity so that it becomes inedible as far as its potential predators are concerned. The black backed oriole in Mexico has however found out that the toxic substances of the butterfly are in its wings and its skins. When it catches one of these butterflies, it simply discards the wings and the skin and eats the rest of it without any consequences to itself.

I cannot see how utter chance can produce all these mechanisms in successive living creatures for the plant, to the caterpillar, to the butterfly, to the bird.

-.-.-.-.-.-.-.-.-.-

Butterflies and moths do not like the hot summer weather in some countries. They prefer to spend the summer away from the direct sun and they sometimes do this by finding suitable cool valleys on higher hills. The butterfly valley on the Greek Island of Rhodes is a very well known such location for millions of butterflies.

Butterflies just like birds use the sun, the moon and the stars to find their way as they search for cooler areas. In New Zealand a huge number of glow worms have gathered together in a huge cave. This cave has an enormous opening facing the route that moths and butterflies normally use on their flight to higher altitudes for a more pleasant environment. The glow worms attach themselves to the ceiling of the cave and create sticky threads that hung from their bodies. At night-time, their tails start to glow. The view that they create is that of the sky with thousands of stars. Flying moths faced with this view get confused and go straight into the cave and then up towards the lights of the tails of the glow worms. Obviously they get stuck into the sticky threads hanging from the tails of the glow worms.

-.-.-.-.-.-.-.-.-.-

Ladybirds haven't got the flying capacity of butterflies. So when the scorching weather arrives in the Central valley California, the ladybirds want to escape and go to other cooler areas. Through some unknown process they know that if they fly straight up to a height of about fifteen hundred meters at that time of the year, there are strong winds that blow Eastwards. They can then a get a free ride for about 80 kilometres to the Sierra Nevada Mountains. When they get there they somehow know exactly into which valley to go and they descend into that valley. Millions upon millions of ladybirds arrive in these valleys. They spend the rest of the summer there as well as the following winter. They return back to Central valley early the following Spring when they know that the winds are blowing in the opposite direction. Again they fly vertically up about fifteen hundred meters and they get a free ride back home.

-.-.-.-.-.-.-.-.-.-.-

The species Empis tessellata normally meet together in mid air before the male takes the female to a suitable twig where they copulate. It is not unusual for the female to eat the male after copulation. To avoid being eaten, the male captures a prey, which he offers to the female as a gift, before copulation. In some cases copulation takes place whilst the male holds onto a twig, whilst at the same time he holds the female and the prey, which the female will eat after the copulation is over. Copulation can last for many minutes or sometimes hours. It is interesting how he can manage to hold himself as well as the female and the prey from the twig for such a long period of time.

-.-.-.-.-.-.-.-.-.-.-

The female of a species of gall wasps which is less than 2mm in length, lays its eggs within a fig which is only about 5 mm in diameter. It enters into the fig through a minute hole at the top of the fig. The hole is so minute that she can sometimes lose her wings as she tries to squeeze into the fig. As she squeezes into the fig, she makes contact with the female part of the flower and deposits the pollen that she carries in her pouches; thus she helps to achieve the desired pollination for the fig. She then lays her eggs. The fig flower immediately envelopes each egg into a gall, which contains food for the young when it hatches. There may be several wasps that enter each fig, but they all die inside the fig soon after laying their eggs.

The males are the first to hatch. As soon as they hatch they start searching for galls containing the females. The male gnaws at the gall containing the female until she comes out and he immediately copulates with her. All the males then get together and start making a tunnel, which the females can use to get out of the fig. The reason that the males undertake this task is because the jaws of the females are not strong enough.

Whilst the males are making the tunnel, the females go round the fig amongst the male fig flowers, which ensures that their pouches are filled with pollen. This is the pollen that will be used at a later stage when she re-enters the fig to pollinate the flower.

When the tunnel is completed the females get out, whilst the blind, wingless males stay behind; they soon die. The young females however, as soon as they get out of the fig are ready to start the same process again of re-entering the fig and pollinating it.

-.-.-.-.-.-.-.-.-.-

The camphor beetle is a tiny creature that lives on the banks of rivers. If it falls into the water it escapes from its predators by a rather ingenious method. It emits a special chemical from its abdomen, which reduces the surface tension at its rear part. The surface tension however is still in full force at the front of the beetle. This produces an imbalance of forces with the outcome that the force at the front drags the beetle along in a forward direction. It is really fascinating to see this little creature being powered along by what some people might describe as a perpetual motion machine, at an apparently phenomenal speed. By moving its abdomen from side to side it can manoeuvre in any direction it wishes.

-.-.-.-.-.-.-.-.-.-

Most insects find it difficult to fly early in the morning before the sun rises because it is too cold for them. This is due to the fact that insects do not have a high body temperature like mammals; they normally take all the heat they need from the sun. The bumblebee with its relatively enormous body needs more power than most bees in order to fly. Yet it has found ways to fly even at quite low temperatures. The way it achieves this is to warm up itself by shivering its wings before take off. The bumblebee has also developed a coat of hair around its body, which enables it to keep a higher temperature.

-.-.-.-.-.-.-.-.-.-

CHAPTER 10

PLANTS AND FLOWERS

Flowers appeared on earth 100 million years ago. They developed nectar to reward insects visiting them in order to transfer pollen. They also developed colours and scent to attract them. Some flower colours can only be seen by insects in the UV region. Bees (insects) and flowers have to coexist and develop together.

Let us now consider some examples from various species of plants and flowers.

A species of orchid has developed a flower that emits a scent to attract an ichneumon wasp. The flower also looks like a female wasp to make the attraction of the male wasp even more effective. In fact the likeness of this flower to a female ichneumon wasp is so striking that the male can copulate with the flower. The pollen, shaped like a horseshoe attaches to the abdomen of the wasp. This orchid depends entirely on the ichneumon wasp for its pollination.

It is difficult to see how random mutations can produce a flower that looks like a wasp with enough likeness to cheat even the wasps themselves. And not simply for the male wasp to be attracted to these flowers, but also to copulate with them. Some people might regard this as science fiction at its best. I certainly was slightly confused when I first heard about this phenomenon.

How can randomness design such a shape? Why didn't the orchid develop a flower the shape of an ant, a bee or a butterfly that did not exist in big numbers in that environment, or were regarded as unsuitable for pollination?

Instead, it designed its flowers to attract a wasp that existed in big numbers in its environment.

One could comment, "Isn't that so clever of the orchid?" The answer is a very affirmative "Yes, indeed"! For the only way that this shape of flower could have been designed is by the orchid itself, with a little help from our Counter RNA, of course.

-.-.-.-.-.-.-.-.-.-.-

A similar trick is played on a thynnid wasp in Australia. The female thynnid wasp spends most of its time digging under trees looking for scarab beetles, which form its main menu. As wings, are a nuisance to digging, she has lost her wings. When it is the right time for her to mate, she climbs up a tree and releases her pheromone. As soon as a fully winged male wasp detects her scent he flies towards her. He takes her from her back and flies off with her, where in mid air they copulate. However the male continues flying with her as if to reward her for her favour to mate with him. He takes her round to various flowers allowing her to have nectar for the first time in her life. This gives her the opportunity to acquire the required nourishment as she will lay her eggs a few days later.

This feature of the thynnid wasp has been exploited by the dragon orchid, which has developed the lip of its flowers to look like the wingless female wasp. It even secretes a scent that closely resembles that of the female pheromone. Above the lip of the flower hang the orchid's anthers and its style. When the male wasp smells the scent and then sees the female body, it flies towards her and as it tries to lift the apparent body of his female mate, the orchid's pollen falls on his back. The next time he is deceived he passes on the pollen to the next mimicking flower.

-.-.-.-.-.-.-.-.-.-.-

Flowers have in some instances developed intricate and elaborate mechanical means to collect and deposit pollen by bees. In one such evolved device the stamen bends over in the form of an arc to deposit the pollen on the bee's back when the bee is detected in the vicinity. Similarly the stigma has developed complex but effective mechanisms for collecting the pollen from the bee's abdomen. These are extremely complicated but well designed mechanisms to effect pollination.

It should be stressed however that one would have expected nature to follow a path of simplicity. Thus the most obvious form of a flower would be circular with a probable large opening on the top to increase the chances of attracting the attention of a bee. It would thus have a conical three-dimensional structure with the narrow part of the cone being necessary for holding the flower to its stem.

-.-.-.-.-.-.-.-.-.-

In another even more sophisticated example, one species of orchid the Coryanthes has part of its lower petal formed into a bucket. It has also created two horns situated above this bucket, which continually secrete small droplets of virtually pure water that steadily drop into this bucket. It has also created a spout halfway up this bucket, so that when the water reaches this level, it overflows through the spout, so that it ensures that the bucket is always half full. The important parts of the petal are situated over the bucket in a hollow cavity with two entrances. This cavity contains fleshy ridges that attract bumblebees. Numerous bumblebees crowed together trying to scrape these ridges, they push each other. When they push each other, some of them fall into the hollow bucket of water below. After they fall into the water, their wings get wet and they cannot fly away, so they are compelled to crawl out of the bucket through the narrow overflow passage. Whilst pushing itself out through this narrow tunnel the bee rubs its back against the viscid stigma and then against the viscid glands of the pollen masses. These get stuck to the back of the bee. When this bee visits another flower and is again pushed into the water, on its way out the pollen on its back is now removed by the viscid stigma, and the flower is accordingly fertilised.

For those that believe that such a contrivance was created by sheer randomness, I think they should take the views of an engineer.

-.-.-.-.-.-.-.-.-.-

It is obvious that the biggest problem that plants face in a desert is the shortage of water. These plants have evolved in such a way as to conserve water and use it as efficiently as possible. They generally have small leaves to minimise the surface area through which evaporation can take place. They frequently drop their leaves during the dry season for the same reason. They take in Carbon Dioxide at night in order to close their pores (stomata) during the day, to reduce evaporation. To overcome the problem of creating enough energy for the plant, the photosynthesis is carried out mainly by the stems of the plant rather than by the leaves – as is the case with other plants.

To be able to soak water from deep-water sources, some of the plants have developed long roots. Other plants have evolved extensive shallow roots, which collect surface moisture from occasional rains or even from heavy dues.

The seeds of some flowering plants can lie dormant in the hot soil for years before rains arrive to help them germinate and grow quickly and bloom for just a few days.

-.-.-.-.-.-.-.-.-.-

The silver sword is a flower that grows only on the crater rim of an extinct volcano in Hawaii. This is a plant that takes many years to actually produce a flower and once it produces a flower, it dies within a few days. The actual flower is one of the most beautiful flowers in nature. Due to the hot environment the plant had to develop means of cooling itself. In common with many other desert plants its leaves allow slow evaporation of water to cool it down. It does this more effectively through some tiny hairs that it has developed on its leaves. The colour of its leaves ensure that they reflect a high proportion of the sun's radiation which again helps to keep the plant cool.

This plant exists mainly only in one place, there are very few of its species, in that place. It reproduces only once after many years. It would not have been possible to evolve as per the Darwinian Theory, which stipulates that:

1. The number of offspring brought into being, is far greater than the number that survive to sexual maturity.
2. Therefore only those that survive can pass on acquired characteristics.

The number of silver sword flowers is so low that it is impossible to apply any form of selection.

-.-.-.-.-.-.-.-.-.-

An equally rare plant that flowers every seven years, producing the biggest flower in the world is the titan arum, which grows only in the central rain forests of Sumatra. It has a spathe, which is about one metre in diameter and its spatix three metres high. The plant grows off a spherical corm, which is about half a metre in diameter, and completely under the ground.

When it first grows from a seed it looks like an ordinary tree with a thin tall stem up to 6 metres high, with three branches at the top carrying small leaves. The corm in the ground grows bigger every year and the leaves wither away and re-grow. After seven years the stem dries up and disappears. Then suddenly the giant bud shoots off the corm from the ground, growing at 20 to 30 cm per day until it reaches its full height.

Its full glory only lasts for two days; the whole flower suddenly collapses after that. The whole structure starts to shrink forming a big watertight bag, with the ovaries of the fertilised female flowers inside it. The stem then suddenly begins to grow taller lifting the enormous pear shaped bag higher.

Soon after, the bag decomposes and thousands of huge reddish berries up to 15cm long drop out. Birds eat these berries and spread the seeds over large areas.

This is one of the few plants that botanists have not been able to reproduce or make it flower anywhere else in any Botanical garden.

-.-.-.-.-.-.-.-.-.-.-

The following examples provide further evidence to support my hypothesis. I present them without any comments.

The mangrove tree is found in swamps and marshes in tropical climates. Its many prop roots assist in binding the mud together, which leads to the formation of land around the roots of the mangrove. To ensure a higher chance of survival of the seeds in a very testing environment, the seeds of the mangrove do not leave the tree as it happens with other seeds. Instead they stay attached to the plant and receive nourishment from it. They receive water, nutrients and food. Even after the seed germinates, the seedling continues to be attached to the parent plant. The parent plant continues the nourishment and in cases where the plant is growing in salty water, the parent plant ensures that the seedling receives salty water in stages so that when it eventually drops into the mud and the salty water, it is used to that environment.

-.-.-.-.-.-.-.-.-.-

The transfer of pollen from one flower to another through insects can sometimes be a hit and miss affair. If a flower allows access to its pollen to all insects, then these insects can spread its pollen to flowers of the wrong species, thus wasting its resources in creating the pollen. Some flowers are actually much cleverer than that. They have developed techniques through which only one species of insect can have access to its pollen. Then this insect becomes the expert on that particular species of flower and tries to limit itself to that flower alone. This ensures a much higher success rate in the transfer of pollen to the right address.

Borage is an annual plant with bright blue pedals. It is used in salads as a herb. The flowers of the borage do not release their pollen to many bees that visit them, even though the bees might actually be touching the stamens of the flower. The anthers of the buzz-pollinated borage flower resonate to a specific high frequency audio sound. This is the sound made by bees of the species such as anthophora, bombus, etc. When these bees visit the borage flower, the frequency of the buzzing of their wings sets in motion the anthers, which resonate at these frequencies and release their pollen grains. The honeybee does not produce the right frequency to cause the anthers to shake off their pollen grains. Studies have indicated that this behaviour of the borage does actually maximize the number of pollen grains carried to other flowers of its species. A mathematically minded flower that analyses its chances of success whilst simultaneously ensuring the minimum wastage of its resources by proper selection of visiting bees!

Horticulturists have copied this technique of the borage and have improved the pollination of potato plants, by artificially vibrating their flowers with sound of the correct frequency using loud speakers.

The pink gentian in South Africa has developed a similar mechanism. Its pollen is held inside a yellow anther, and it can only be accessed through a small hole near the top of the anther. The carpenter bee is the only insect that is allowed to have access to this pollen. When the bee reaches the anther it reduces the speed of the beating of its wings, so that the sound they make also changes. As the frequency reaches that of middle C, which appears to be the resonance frequency of the anther, the anther immediately starts to vibrate and the pollen is emitted through the small hole, like water from a tap.

-.-.-.-.-.-.-.-.-.-

Leguminous flowers have developed a way to fasten the entrance to any visitors except to bees that have the intelligence and strength to open it. This is a complex device that incorporates a spring-loaded mechanism. To open the entrance the bee has to apply its weight onto the platform of the flower. It then pushes its head against the upper petal. Simultaneously its middle legs on the keel and its hind legs push against the two wing petals. Then presto! The way is now clear for the bee to access the nectar with its proboscis and of course for its head to make contact with the style and stamens.

Some pea flowers are even more discriminatory than the leguminous flowers. They only allow some species of bees to enter whilst they reject others. In this case the bee's brow is pushed against the standard, while the fore legs and the middle legs are resting on the wing petals, and the hind legs are pushing down on the keel.

The incredible thing is that bees that know the code to each flower do not get confused. They remember the code for each flower and use it accordingly. It is like a person that carries many keys and knows which key to use for each lock. Bees never hesitate which code to use. One can tell this by the speed with which they enter a flower once they get to it.

-.-.-.-.-.-.-.-.-.-

The common vetch Vicia sativa has developed a means by which it produces a small amount of nectar outside the flower itself. This nectar is actually at the base of the flower stem. This nectar is not for the benefit of pollinating bees but for ants, which guard the flower nectaries from other potential thieves that might also damage the flower. The ants themselves never try to enter the flowers.

-.-.-.-.-.-.-.-.-.-

The birdcage plant, which is found in the American deserts, needs long roots to go deep enough to search for water. The wind in the desert does however sometimes blow away the sand around the roots of the birdcage exposing them to the sun. The birdcage dies soon after. However as soon as the plant dies, its stems curl in such a way as to form a light hollow ball, the size of a tennis ball. This light ball also contains seeds of the plant. As it is so light, it is blown by the wind. It can travel long distances sometimes many kilometres over different terrains, until it eventually finds itself into a deep where the wind cannot blow it away. At this stationary position, sheltered from the wind, the seedpods open and the seeds germinate. They now have a much better chance of survival than in the previous exposed location. The seeds carry enough nutrients to ensure a good start, until its roots reach deep into the ground to find water.

-.-.-.-.-.-.-.-.-.-

The seeds of a tree in Brazil, the monkey's dinner bell, are transferred through an exploding pod. The large pod containing the seeds is warmed on one of its sides by the sun, whilst the other side remains cool. The moisture from the warmer side evaporates quickly, so that, the warmer side dries up. The drying up process creates a mechanical tension in the pod, which results in the explosion of the pod. This explosion of the pods can be quite loud and sounds like bullets being fired in the forest. Following the explosion, the seeds can be thrown many metres away.

-.-.-.-.-.-.-.-.-.-

All plants have to produce their own food for their survival and growth. This food consists of starches and sugars that are manufactured by the plants using their own chlorophyll, which is abundant on all leaves. The process of manufacturing their food is called photosynthesis and apart from chlorophyll, it also requires sun, carbon dioxide, water and some minerals contained in the water, which is absorbed through the roots.

Without sunshine photosynthesis cannot take place. In the Tropical forests where the trees can be very tall and growing very close to each other, there is very little sunlight reaching the shorter plants close to the ground. To overcome the problem of collecting enough light some of these plants have grown large leaves, some of which are well in excess of three square metres.

Other plants have developed different techniques to get as much light as possible for their requirements. Some of them have developed a coat of red or purple pigment on the lower side of the leaves, which reflects any light that has not been absorbed by the leaf and was otherwise going to go through the leaf.

Another group of plants have developed a transparent layer on the top side of their leaves that act as very small lenses that focus the light onto the grains of chlorophyll on the leaf.

-.-.-.-.-.-.-.-.-.-

The marsh pitcher, which grows on very poor soils in Venezuela, takes its nourishment from dead insects that it manages to attract into its leaves that are shaped like a cone and are half filled with water. The cones have an overflow device that ensures that they are never filled with water all the way to the top. They get the water from the rainfall, which is fairly abundant in that region.

The pitcher produces nectar with a very attractive scent for insects. As the insects go into the conical leaves looking for nectar they tend to slide down slowly, as the pitcher has developed hairs on the internal surface of the cone that point downwards and are very slippery. The insects eventually find themselves floating in the water and unable to climb out of it.

Bacteria eventually ensure that these insects turn the water into a nourishing soup for the plant, which it absorbs.

There are other species of pitchers around the world that have modified the above mechanism and in some cases they have made it even more complex, always trying to improve their effectiveness to attract more and more insects.

-.-.-.-.-.-.-.-.-.-.-

The Venus flytrap is probably the most well known insectivorous plant. It grows mainly in North and South Carolina. Its leaves are made up of two lobes that are hinged together. The lobes are completely surrounded by sharp spikes. The internal surface of the lobes carries two or three sensitive hairs, which give a signal to the leaf when an insect touches them. The lobes do not act after a single signal from the sensitive hairs. In order to act there must be two successive signals within a period of twenty seconds. The reason for this is assumed to be that one single signal may be due to something that is not a moving insect.

As soon as the condition of the two signals within a period of twenty seconds is satisfied, the lobes immediately close like a mousetrap. It is estimated that they snap within one third of a second. The insect is caught within the closed lobes, which immediately start squeezing it. The flytrap then starts secreting from its glands digestive solutions including hydrochloric acid, which start to dissolve the insect's body, making it ready for the flytrap to absorb it.

One important aspect of this mechanism, apart from the fact that it needs two signals to activate the lobes, which ensures that it is dealing with an insect and not for example a small leaf, it also ensures that it is dealing with a fairly sizeable prey. If the insect that activated the mechanism is a very small insect, then it allows it to escape, before the two lobes start the squeezing. This is because the flytrap realizes that the amount of nourishment it will receive from the small insect is not going to be recovered, bearing in mind the energy that it requires to replenish the amount of dissolving juices required to break up the small insect.

-.-.-.-.-.-.-.-.-.-

To avoid its bird predators, a moth caterpillar in the Tropical forests of Borneo has developed a technique by which it creates a cover for itself so that the birds cannot see it whilst it carries on eating the leaf below. To create this cover, the caterpillar eats away two semi circular areas of the leaf. It then spins threads of silk across the two loose segments of the two semi circular areas. As the thread dries it pulls the segment of the leaf on top of the caterpillar. Even though this process takes more than two hours, the caterpillar can enjoy a meal in peace.

A European moth lays it eggs in twigs. The young caterpillars spin a large covering for themselves, big enough to cover them all. They use this hide out to conceal themselves from predators during the day, whilst at night they form a procession of several hundred to attack leaves, which are sometimes a long distance from their hide out. Every day one member of the group is sent out on a mission to explore the area for more leaves. This explorer leaves a trail of scent behind it, which is given off from glands at its rear end. The explorer uses this trail to find its way back. The following night the group of caterpillars looks for the double trace of the explorer. They smell the scent left behind. From the scent they can tell whether the explorer found leaves to eat. If it did not find any leaves they do not bother following that trail. Furthermore if there is only one trail it means that the explorer had met with predators on its way, so they definitely avoid following that trail.

-.-.-.-.-.-.-.-.-.-

To keep browsing animals away from their leaves, some plants have developed various techniques by which they discourage animals from getting too close to their leaves. The hawthorn has developed spikes on its twigs. Most animals would not even touch a hawthorn. In the tropics, palm trees have developed needles that are over 20cm long. Some animals have developed long and tough tongues to overcome this problem. But this again is the problem of techniques used by the hunted and the hunter described elsewhere in the book.

Other plants such as the nettle have developed glassy needles that when touched, the needle penetrates the skin and it breaks off from the leaf of the nettle. This can cause a cut on the skin. Furthermore the needle ejects a poison onto the wound that can cause considerable pain. There are other plants that have developed similar stinging needles that are much more venomous that those of the ordinary nettle.

Some plants that have no poisonous needles have developed the appearance of plants that have them. This is mimicking as we get in animals that has been discussed in other parts of this book. For example the deadnettle has leaves, which appear to be very similar to those of the ordinary stinging nettle.

-.-.-.-.-.-.-.-.-.-

Plants always try to find ways to defend themselves against animals. Apart from growing thorns on its branches, the African acacia has also developed a mechanism through which it emits distasteful or even poisonous chemicals to keep animals away. The plant creates this chemical as soon as an animal starts eating its leaves. The most incredible thing about this mechanism however is that as the plant creates the obnoxious chemical, it also emits ethylene gas to warn its neighbours that it is being attacked and it is therefore creating the obnoxious chemical. As soon as the neighbours smell the ethylene gas they realize that this is a message that one of the acacias is being attacked; they immediately start manufacturing the obnoxious chemical and simultaneously emit ethylene to warn other acacias further down the line.

The best example of mimicry in plants is probably that of the passion vine of South America. The leaves of this particular plant are the favourite meal of the caterpillars of the black, yellow and red butterfly. The butterfly always lays its eggs on the leaves of this vine to ensure that the young caterpillars have plenty to eat when they hatch. When the butterfly reaches a leaf to lay her eggs, if she sees that another butterfly has already laid her eggs on that leaf, then she looks for another leaf. This is to ensure that her young caterpillars do not have to share the leaf with other caterpillars.

The vine has taken advantage of this habit of the butterfly and has developed a feature whereby it grows small yellow knobs on some of the leaves that look like the eggs of the butterfly. Thus these leaves are avoided by the butterflies and are therefore saved from the caterpillars.

It is thought that the vine is clever enough not to grow these knobs on all its leaves, as the fly would be unable to find somewhere to lay its eggs and would therefore be forced into a situation where it has to develop a counter mechanism by which it would detect the knobs from the true eggs. Does this sound to you as if such a counter mechanism could be developed through randomness?

Another interesting example of plant mimicry is that of the pebble plants in the African deserts. They have only two leaves that are very succulent and tender and most animals would love to make a meal of them. But they don't because they do not know that these leaves are there. These leaves grow amongst desert pebbles and always take the appearance of the pebbles around them.

-.-.-.-.-.-.-.-.-.-.-

The rattan is a kind of palm tree, whose stems grow a bud at their tip. The whole stem can die if the tip is damaged. Many animals like to eat this juicy succulent tip. The tip is covered by very sharp spines, which keep away most predators. There is however another defending mechanism, which ensures that predators don't get near the rattan.

Within the brown husks that are always around the main stem, a colony of ants create a big cavity, which they convert to their nest. As soon as a predator approaches, the ants in the nearest husk immediately start beating their heads on the husk in unison. This produces a surprising loud noise. The other ants in the other husks when they hear their brothers thumping their heads, they also start the same procedure. They all act in unison. Within a few seconds the whole plant appears to be an enormous source of sound. Soon after some of the ants come out of their nests and encircle the stem. They try to show to any potential predators that they risk being beaten if they touch the tip of the rattan. Apart from providing a home, the rattan returns the favour to the ants in an indirect way. It allows small aphids to suck the rattan's sap. These aphids excrete drops of honeydew, which the ants drink. The aphids are actually used like a herd by the ants, just like humans keep herds of sheep or cattle for their milk.

-.-.-.-.-.-.-.-.-.-.-

We have seen above how acacias use the emission of chemicals to defend themselves against predators. Acacias have also developed long thorns that look like the horns of some animals, to stop large animals from getting to their leaves. There are insects however that are leaf eaters and can also cause a lot of damage to these thorns and to the whole plant. Ants have once again come to the rescue of plants. Ants create their nests inside the hollow of the horns. They come out regularly to patrol the acacia branches looking for any insects, which they attack immediately and take them to their nests. Apart from providing a home for the ants, the tree rewards the ants even further by providing them with nectar. Glands, situated near the lower part of the leaf stalks, produce this nectar. The plant also grows some orange beads on the tips of the leaflets, which contain fats, and which the ants eat.

The ants actually do a bit more for the plant. They do regular surveys on the ground around the plant and destroy any seedlings that may have appeared and could eventually compete for space with their host. They also ensure that if any other plants extend their branches into the branches of their host, they attack these branches and destroy their leaves. As the colony of ants grows they expand their territory into other thorns. Unfortunately there may be more than one colony of ants on the same tree. This results in serious battles and many ants are killed during these battles. Usually the winners are the colony with the highest number of soldiers.

-.-.-.-.-.-.-.-.-

EPILOGUE

I have tried in these few pages to present a new hypothesis to elucidate some observations that could not be explained in any other way.

I supported my hypothesis through various scientific principles as well as with nearly two hundred examples from nature.

However as I explained repeatedly in my book, I expect that the clique of Darwinists will discard my hypothesis, without even bothering to go beyond page one of my book. It is a great pity that Scientists throughout the ages have demonstrated that they can be as dogmatic about their beliefs as much as politicians, economists or statisticians can be about theirs. And this in spite of overwhelming evidence that established scientific theories can sometimes be proven wrong or at the very least modified to accommodate further observations.

I sincerely hope that somewhere out there, there will appear somebody that will study my hypothesis in its true perspective and become its champion. Just like Darwin needed the support of people like Huxley, I now need the support of someone to convince people of the necessity to look into my hypothesis more closely and possibly carry out the necessary scientific work to provide the absolute truth for it.

As I said in my introduction, I have been planning this book for over thirty years, mainly because I wanted to present it in a simple form, just like Darwin did with his "Origin of Species". The biggest obstacle however was for me to get enough courage to fight all those that would condemn me as a heretic or a crank.

However one of the driving forces that eventually convinced me to go ahead was the fact that possibly some scientists will realize that **my hypothesis does not contradict Darwinian Theory, it simply complements and augments it.**

Another reason for going ahead with it was the fact that through the ages there have been numerous examples of Scientists who were treated as heretics when they came up with something new. Galileo is probably the best example that comes to people's minds. However a better example in the field of Genetics is the man that actually founded Genetics, Gregor Mendel. Mendel did such a fantastic scientific work on the acquired characteristics of peas but his work was pushed aside as of no significance. Scientists at the time did not show any respect for his work, mainly because he was not regarded as a scientist. Mendel was a monk in a monastery in Austria. A Dutch Scientist eventually brought his work to light, many years after Mendel's death.

I sincerely hope that Science will be fairer to me than it has been to Mendel.

APPENDIX 1

I would like now to return to the controversial proposal of Steele and Gorczynski discussed earlier in chapter 3 and who have demonstrated that it is possible for mice to acquire immunity from some diseases, which were acquired by either of their parents during their own life time.

I want to take these findings a step further. I would like to consider some examples where this observed DNA modification could possibly explain some observed every day phenomena. I am not going to be dogmatic on the examples provided below; the reason I provide these examples is simply because if eventually they prove to be correct, the implications are frightening to say the least.

As a first example I would consider the rather contentious issue of the acquired characteristics of skill from generation to generation. It has been suggested in numerous writings that the blacksmith acquires his large muscles not through genetics but through the continuous use of his hammer. It is also obvious that only those that are physically strong enough to take up this particular profession would do so. This is not a job for weaklings. Thus a blacksmith must have large muscles before he starts his profession but whilst exercising his profession, his muscles become continuously bigger.

To carry out this profession however, one should have the right ability or skill or be in a position to be trained to be able to achieve a reasonable level of expertise to carry out this rather unique profession.

It is normally accepted that the fact that the son of a blacksmith follows his father into his profession is not because he genetically acquired his father's skill in the profession, but because he genetically acquired his father's physique and muscles and is thus able to carry out that profession. As mentioned earlier, it is said that even on his death bed Darwin was saying repeatedly that the blacksmith's son must have acquired his big muscles from his father.

Let us now look at the ability of people to play football. To be able to achieve the highest standards of playing football, for example to become a professional footballer there are some essential physiological requirements. To be able to kick the football hard, one needs to have strong calf muscles plus strong foot and toe muscles. To be able to kick the ball accurately in a particular direction, one has to have very good control and agile movements through the ankle. To be able to dribble the ball past an opponent requires the right shape of foot and toes as well as great dexterity to move these at precisely the right time and the right angle to achieve the best possible kick to place the ball in such a position and direction that would confuse the opposing footballer. Obviously this requires full control of such motions by a brain that reacts extremely quickly to the continuously changing position of the ball especially when the opposing player touches the ball. A good footballer must have the right physique to be able to withstand the pushing by opposing players and the ability to regain his balance once pushed. He must also have above average size of lungs to be able to store large quantities of oxygen to keep him going at times of high demands of oxygen by the body. Like most other athletes, a footballer must be able to produce the right level of metabolic rate in order to create the amount of energy required, especially in the situation where he has to run very fast over short distances.

In addition to the above requirements a footballer must also have speed and stamina. Heading the ball accurately and powerfully can also be of great advantage. This requires him to have powerful and controllable muscles in the neck as well as a strong skull.

He must also have the ability to plan what to do next as far as his position relative to the rest of the team is concerned and intuition as to which should be his next movement and position in the field relative to the positions of all the other players of both teams. A good footballer can sometimes change the outcome of a match through a good pass to one of his team mates, or even lose a match for his team through a wrongly timed tackle against one of his opponents.

Commitment of a footballer in a tackle against another footballer is of obvious advantage even though it is well known that many great footballers find alternative ways to achieve the required effect without embarking on an apparently brutal clash.

It has to be pointed out that one of the reasons that football is such a popular sport around the world is because a good footballer does not have to be an outstanding athlete. A top athlete has different requirements and his physique must be much more responsive to his particular sport than that of a footballer. The footballer obviously must be a good athlete but not necessarily a top one.

So bearing in mind that a footballer must have so many unique abilities, many of which are genetically acquired, it is not perhaps surprising to see so many successful football players being closely related within a family.

In addition to the basic physical attributes, a boy must have the will and desire to train systematically to achieve a high level of skill in football. Most boys however with the right physical attributes dedicate themselves to football from a very young age. The game is so popular that not many children would turn down the opportunity of playing football.

If we take England as an example there are around 100 top professional teams, with an average of about 25 professional players. Thus at any one time there are about 2,500 professional players in England. If we assume that the professional career of a footballer is around ten years, then we could conclude that we have around 250 players who become professional footballers every year. From these figures we could then make a further assumption that every year there are about 250 boys born that are destined to become professional footballers.

The total number of boys born in England every year is approximately 350,000. Thus the chances of a boy becoming a professional footballer are only about 7 in 10,000, or one boy in every 1,400 boys born.

With such a low probability of a boy becoming a professional footballer is it not rather surprising that we frequently have two brothers becoming professional footballers? Even more so when two twins become professional footballers! In fact we have frequently seen such occasions where brothers or even twins become international footballers, where the probability of such an event is so low that one would not even contemplate using the word randomness into such an argument. The chances of this happening on a random basis are probably similar to a house being struck twice by lightning.

What is also of significance is that frequently we get the situation where sons or grandsons of a professional footballer, follow their fathers as professional footballers. This is not only because they grow up in an environment where football is the livelihood of the family. Unless a young boy has the right skills, as discussed above, he will not make it to the top. Thus we can conclude that there are many attributes required for a professional footballer that are genetically acquired.

As a further example of this genetic characteristic one could site the case of the Asians in the United Kingdom. There are around 3 million Asians (mainly Indians and Pakistanis) in the United Kingdom which is around 5% of the total population. Yet there is hardly one top Asian footballer in the UK even though most of the young British Asians are third generation. One should mention here that traditionally Asians follow cricket much more than they follow football but I think within the next few generations there will be as many top Asian footballers as the rest of the population. However a visit to numerous parks around the Country on Sunday mornings would show thousands of young Asians participating in football admittedly at a relatively low level of competence.

As I said above the physical qualities of a successful athlete are different to those of a successful footballer. It is rather strange that virtually all the renowned African long distance Olympic and World champions were born in a few well defined areas at high altitudes in Kenya and Ethiopia. It has also been pointed out by researches at the Department of Extreme Endurance in Glasgow University that many of these athletes are close family relatives. One other aspect that they have pointed out is that these athletes were running tens of kilometres every day from quite a young age to go to school. It appears as if there is no apparent physiological feature that is similar to other high altitude people such as those of the Himalayas or the Andes. However the fact that such a feature has not yet been discovered does not necessarily mean that it does not exist. But as I mentioned earlier in this book the Counter RNA does not always help in providing identical solutions in every occasion. A feature could exist that makes these highly successful athletes to be so unique in their achievements.

However one should also consider the possibility of a Transcription factor as discussed in Chapter 3, as being responsible for these outstanding achievements. Thus there may not be a genetic reason for their achievements. Their achievements may be because they were brought up to run these enormous distances from childhood – just like the toughening of the sole of the foot as discussed in Chapter 3.

This now introduces a new phenomenon: The possibility that the achievements of some athletes (such as footballers) may be genetic whilst of others (long distance runners) may not be genetic at all. This could be because a successful footballer needs to have numerous features some of which may be genetic and some not. This may also apply in the case of long distance runners, but probably not to the same extent.

-.-.-.-.-.-.-.-.-.-

I am now going to go a step further. I am going to venture into a sometimes highly disputed and controversial area. This is the area of human behaviour and reflex action. Many people dispute as to whether reactions and reflexes are acquired genetically. But bearing in mind that some of these are definitely acquired genetically then there is no reason as to why others should be rejected as not acquired genetically. For example the suckling of milk by all mammalian offspring from their mother's breast has to be an acquired feature. The newborn must have inherited such a feature otherwise it would not have been able to survive.

If we therefore assume that such a characteristic is genetically acquired then one could assume that other reflexes and reactions could also be genetically acquired. Let us now take one of these reflex reactions concerning humans. If we were to throw an object, say an apple to a person expecting that person to catch it whilst sitting on a chair, that person would raise his or her arms and getting them together whilst opening their hands. If that person were a man or boy, the person would instinctively close his legs together whilst trying to catch the apple with his open hands. If that person however happens to be a woman or a girl then she would open her legs apart whilst trying to catch the apple with her hands. This reflex reaction is actually described in a classic English Novel. But I have tried it many times myself and proved that this is definitely statistically true based on a number of throws with different people.

One would wonder why this is the case. One explanation would be the fact that women and girls who normally wear dresses or skirts (at least they used to until recently) would open their legs so that if they fail to catch the apple it would fall into their dress under their hands.

For men and boys however who wear trousers they would have to close their legs to catch the apple if they fail to catch it in their hands, for if they do not close their legs the apple would simply roll of their legs and onto the floor.

The implications of this reflex reaction are enormous, for such a reaction could only have developed within the human race during the last two or three thousand years. For previous to that both men and women were wearing animal skins, not dresses and trousers. These were invented very recently as far as human evolution is concerned.

This is another example where Darwinian Theory fails to give any explanation whatsoever.

-.-.-.-.-.-.-.-.-.-

Let us now look at the evolution of languages in different regions around the world. Look at the following names of places, rivers, etc.

Zaire, Zambezi, Zambia, Zanzibar, Zimbabwe, Zomba, Zulu, Tanzania, Mozambique, Zaramu, Zastron,

Zawi, Zimbwe, Zimba, Zongo, Zumbo, Swaziland, Brazzaville, Mwanza.

These are all names of Countries, places, rivers etc in the Southern region of Africa. I think the frequency of the letter "Z" is rather obvious for that region of the world. In fact 5 of the 8 countries in that region have the letter "Z" in their name – South Africa, Angola and Malawi being the exception. But then these names are not African anyway.

The fact is that the letter "Z" is for one reason or another very common in that part of the world. In the English language fewer than one in a thousand words start with the letter "Z". In the Collins Family Dictionary with 60,000 references only 59 start with the letter "Z", i.e. fewer than one in a thousand. In the Geek Dictionary of Babiniotis there are 370 Greek words starting with the Letter "Z" in a total of 150,000 references i.e. fewer than 2.5 in a thousand.

The question is why is the letter "Z" so common in these countries?

Obviously the people in these regions must find it easy to pronounce the letter "Z". The structure of their larynx, their vocal cords, their throat and the cavities of their mouths is such as to find it easy to pronounce the letter "Z".

This region is not unique in the world where there is a high frequency of usage of this letter in their language. Russian, Polish, German, Hungarian, Dutch, Chinese and others also have a high proportion of words containing the letter "Z", even though there may be variations in the actual pronunciation of the letter "Z" for each language. In fact on the Hungarian computer keyboard the letter "Z" has been moved to the position of the "Y" of an English keyboard for easier access, because it is so common.

My premise is that this was an evolutionary trend that depended upon anatomical reasons – the structure of all those organs responsible for the production of sound and speech. It is quite possible that the diet of the people in various regions modifies the shape and size of these extremely sensitive organs to respond better to certain sounds than others. As there are differences between the actual pronunciations of the letter "Z" in the above languages I do not necessarily expect that there would be Anatomical similarities between people of these different Countries.

As further examples I can quote Arabic which to non-Arabs it appears as if the sound "hala" dominates any conversation in Arabic.

I could also quote a number of languages where the letter "A" dominates. I would like to ask you a quiz. Name a Country where the total number of "A's" in the name of the Country, the name of its capital and the name of its President combined is 14.

When I put this question to a number of people, the favourite answer was Sri Lanka, because it is well known that many Sri Lankan names do actually contain a large number of "A's".

The answer to the above quiz however is Madagascar. The name of the current (2007) President of Madagascar is Marc Ravalomanana and the name of its capital is Antananarivo. This is 14 "A's" in a total of 38 letters.

I have to reiterate that the examples of "rapid evolution" given above do not form part of my convictions and are not given as proof of my hypothesis. The only reason I am referring to them is because of the findings of the two groups of researchers mentioned above and my duty as a Scientist not to hide any views that may act as a catalyst for further investigations by other Scientists who might be in a better position than me to carry out such work.

APPENDIX 2

Table showing number of oak trees assuming one acorn per tree per year becomes a mature tree. The number of acorns is based on trees starting at 10 years old. Normally oak trees start producing acorns at 3-5 years old. A fully mature tree can produce up to 100,000 acorns. Oak trees can live in excess of 100 years,

10 years	10 trees
20 years	65 trees
50 years	16,687 trees
100 years	136,927,350 trees
150 years	1.09682E12 trees
156 years	3.30938E12 trees

Assuming the total ground area covered by each tree to be about 200 m2 the total area covered by all these trees is
$$0.66187E15 \text{ m2}$$

Assuming radius of the earth	12,756,000 km
Total surface area of the earth	2.025E15 m2
Land surface (30% of total)	0.6075E15 m2

Thus the whole of the surface of the earth would be covered with oak trees in 156 years if only one of the 100,000 acorns from each tree per year would become a mature tree. Yet we know that the earth's surface is not covered with oak trees. Therefore the success rate of acorns resulting to trees is less than 1 in 100,000. This is exactly what Darwinian Natural Selection is all about.

Thus Darwinian Natural Selection is fine the way Darwin described it, i.e. the number born within a species is much greater than those that survive to sexual maturity.

APPENDIX 3

The following table shows the theoretical cumulative number of female rabbits within 8 years starting with one pair of rabbits. I have assumed in these calculations 4 litters per year (normally rabbits can have 6-8 litters per year). I also assumed that 2 females are born per litter. The total number of rabbits per litter is normally 5-10. I also assumed sexual maturity of six months and a gestation period of one month. Rabbits can live up to 10 years.

1 year	6 rabbits
2 years	106 rabbits
3 years	890 rabbits
4 years	7,390 rabbits
5 years	60,954 rabbits
6 years	503,914 rabbits
7 years	4,165,562 rabbits
8 years	34,434,048 rabbits

Obviously these are not the numbers we will get in practice. That is why Natural Selection can be applied in this case.

www.ingramcontent.com/pod-product-compliance
Lightning Source LLC
Chambersburg PA
CBHW060828170526
45158CB00001B/113